THE ANCIENT PINEWOODS OF SCOTLAND
A COMPANION GUIDE

CLIFTON BAIN

With drawings by Darren Rees

THE ANCIENT PINEWOODS OF SCOTLAND: A Companion Guide
CLIFTON BAIN with drawings by Darren Rees

First published in Great Britain and Ireland in 2016.
Sandstone Press Ltd
PO Box 5725
One High Street
Dingwall
Ross-shire
IV15 9WJ
Scotland

www.sandstonepress.com

All rights reserved.
No part of this publication may be reproduced, stored or transmitted in any form without the express written permission of the publisher.

© Clifton Bain 2016
© All drawings Darren Rees 2013
© Images as ascribed
© All maps RSPB 2013. Contain Ordnance Survey data © Crown copyright and database right 2012 and using data from the Caledonian Pinewood Inventory, Forestry Commission Scotland 1998.

Editor: Robert Davidson

The moral right of Clifton Bain to be recognised as the author of this work has been asserted in accordance with the Copyright, Design, and Patent Act, 1988.

The publisher acknowledges support from Creative Scotland towards publication of this volume.

ISBN: 978-1-910124-92-5

Book design and cover by Heather Macpherson at Raspberry Creative Type, Edinburgh
Printed and bound by Bell and Bain Ltd, Glasgow

ACKNOWLEDGEMENTS

In this companion guide to the original 'Ancient Pinewoods of Scotland – A Traveller's Guide' I wish to thank again all those individuals who helped make this project happen. My colleagues and friends at RSPB, Scottish Natural Heritage, Forestry Commission Scotland and the Scottish Wildlife Trust have been a constant support and share my desire to introduce a wider audience to the joy of these ancient woodlands.

Thank you to Robert Davidson at Sandstone Press, Heather Macpherson (Raspberry Creative Type) for design and Darren Rees for artwork.

Gaelic translations were made with the help of Jacob King, Ainmean Àite na h-Alba and Gavin Parsons, University of the Highlands and Islands.

Photo Credits: Photography by the author except for the following:

Lorne Gill/SNH – page 14 (red squirrel), page 62 (Loch Maree), page 86 (Glen Affric), page 88 (Glen Affric)

Desmond Dugan – page 21 (red deer), page 29 (Abernethy)

Laurie Campbell – page 7 (Capercaille), page 82 (Glen Cannich), page 180 (Loch Tulla)

Jon Mercer – page 118 (Glen Loy)

Paul Chapman – page 174 (Glen Ferrick), page 176 (the Finlets)

Syd House – page 194 (Glen Falloch)

Richard Johnstone – endpiece (the author in Ballochbuie)

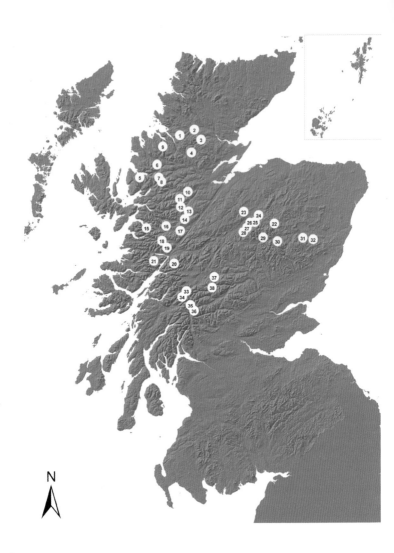

CONTENTS

Acknowledgements	3
Contents	5
Introduction	9
The Ancient Pinewoods of Scotland	9
Wildlife	12
Early history of the pinewoods	16
Prehistoric Pines	17
The Medieval Forests	18
The 17th century	18
Survivors of the Jacobite Uprising	19
The enemy within	20
Pinewoods from the middle 20th century to the present day	22
Climate Change	24
Looking to the future	25
Visiting the woods	27
Using this guide	32
Key to maps	32
The Northern Group	35
1 Rhidorroch	36
2 Glen Einig	40
3 Amat	44
4 Strath Vaich	48
The Western Group	53
5 Shieldaig	54
6 Loch Maree	58
7 Coulin	64
8 Achnashellach	68
9 Coir' a' Ghamhna	72
Strathglass Group	77
10 Glen Strathfarrar	78
11 Glen Cannich	82
12 Glen Affric	86
13 Guisachan and Cougie	90
Great Glen Group	95
14 Glen Moriston	96
15 Barisdale	102
16 Glen Loyne	106
17 Glen Garry	110
18 Loch Arkaig and Glen Mallie	114
19 Glen Loy – Coille Phuiteachain	118
20 Glen Nevis	122
21 Ardgour	126

Strathspey Group	131	32 Glen Ferrick and the Finlets	174
22 Glen Avon	132		
23 Dulnain	136	**Southern Group**	179
24 Abernethy	140	33 Black Mount	180
25 Glenmore	144	34 Glen Orchy	184
26 Rothiemurchus	148	35 Tyndrum	188
27 Invereshie and Inshriach	152	36 Glen Falloch	192
28 Glen Feshie	156	37 Black Wood of Rannoch	196
		38 Meggernie	200
Deeside Group	161		
29 Mar	162	Safety and access	204
30 Ballochbuie	166	Bibliography	205
31 Glen Tanar	170	Useful websites	206

Male capercaillie displaying in Abernethy Forest Reserve

INTRODUCTION

It is surprising how few people are aware of the ancient pinewood remnants, tucked away in the far corners of Scotland's remote glens; survivors of woodland that cloaked much of the land several thousand years ago. They have not gone completely unnoticed however. Queen Victoria mentioned the splendid beauty of the 'ancient fir woods' in her Highland Journal and in present times, pine trees adorn tourist brochures as iconic images of Scotland.

An essential characteristic of these ancient pinewoods is that the trees have naturally seeded and grown to maturity, repeating the cycle of life and death over thousands of years. Spectacular wildlife has found refuge in this diverse and mature habitat. Past changes in climate and human activity have taken their toll on the original tree cover. Most notable over the last 200 years has been the huge increase in managed herds of sheep and deer grazing on young trees. In some places the oldest trees are nearing the end of their life without a single sapling having survived to replace them. Fortunately, dedicated effort has led to great improvements in the management and legal protection of the Caledonian pinewoods (as they are officially named under wildlife law) helping turn their fate around. This is one of the great success stories in nature conservation but there is still a challenge ahead in managing this vulnerable and important habitat.

The Ancient Pinewoods of Scotland

Scots pine is the common name for a species of tree (*Pinus sylvestris* L.) which is characterised by its flaky, reddish brown bark and widely found around the world from the Atlantic to the Pacific. The natural range of this magnificent plant stretches latitudinally from Norway to Spain and in longitude from west coast Scotland across Europe and Asia to Siberia and north east China. Scots pine has survived in Europe for at least 1.6 million years. In Scotland, the pines are of a unique variety (*Pinus sylvestris* var *scotica*) which has adapted to live in more wet and windy conditions.

Pinewood regeneration, Abernethy Forest

Individual trees can live for as long as 600 years but most survive for around 250. Some of the oldest recorded living pines in Scotland are in remote Glen Loyne where, in the late 1990s, scientists estimated one to be 550 years old; having started its life in medieval times, when Scotland's King James II was crowned and King Richard III of England was born. Ancient specimens can also be found in more accessible sites such as Ballochbuie and Rothiemurchus. Although slightly younger at around 350 years, these spectacular trees were seedlings when Oliver Cromwell invaded Scotland.

In Scotland today, the terms *ancient* and *natural* are used to describe woodlands which have originated before 1750, without human planting or influence from cultural activities such as timber felling and livestock grazing. Since most woods in Britain have had at least some human impact the term 'ancient semi-natural' is sometimes used. Scots pine woods are also described as 'native', as they support typical British species of trees which have grown here since the last Ice Age and which were not introduced by humans. Today, new native pinewoods are being established to help recover some of the past losses. If managed sensitively, these will become the ancient pinewoods of the future.

Within Britain, the natural distribution of Scots pinewoods is restricted to the Scottish Highlands. Of the surviving ancient pinewoods, the southernmost is a sparse group of trees in Glen Falloch, near the north end of Loch Lomond. At the western edge of the Scots pines' entire world range sits Shieldaig, near the Atlantic coast, beyond the remote Torridon mountains. The Deeside forests of Glen Tanar and Birse are at the eastern limit in Scotland and at the northern edge is Glen Einig in Sutherland, although small groups of pine to the south of Ben Hope may be the most northerly individual trees. It is also worth mentioning that in England and Ireland there are a few individuals growing in remote peatlands, which local naturalists passionately believe to be descendants of the long gone pinewoods.

One of the most compelling features of the ancient pinewoods, compared to commercial plantations, is their great diversity of structure, with irregular spaced trees of all shapes and sizes reflecting the different soils, slopes and water levels. This rich mosaic of ground conditions is one of the consequences of Scotland's climate and geology. On well-drained sandy soils, tight groups of naturally occurring, straight, tall trees can grow to twenty five metres whereas on wet patches of boggy ground, stunted, twisted trees often grow no higher than two metres, even after a hundred years. Scots pines are able to survive in the thinnest of soils on rocky crags and steep riverside gorges, where their exposed roots and stems wind round boulders like the wiry, strong limbs of

Scots Pine (Pinus sylvestris)

a rock climber. All of this diversity contrasts with the timber plantations, where ground conditions are made more uniform by ploughing before the trees are planted.

Individual Scots pine trees come in different forms including wide and bushy types, straight stemmed with an umbrella-like crown and others whose branches decrease in size up the tree to create a conical shape. Within the forks of some multiple stemmed trees it is common to find plants, and even other small trees growing. Pine can survive in the harshest of conditions from the salt-laden sea winds at Shieldaig to the steep mountain slopes of the Cairngorms, where stunted pines grow at 900 m.

Birchwood with pine on Loch Garry

Broadleaved trees are an important part of Scotland's native forests with birch (*Betula* species) most often found mixed with the pines, along with oak (*Quercus* species), rowan (*Sorbus aucuparia*), juniper (*Juniperus communis*) bushes and other shrubs. Aspen (*Populus tremula*), which is commonly found in Scandinavian pinewoods is not as abundant in the Scottish woods but small groups can be found in the north and west, their distinctive trembling leaves turning brilliant yellow in autumn.

Wildlife

The wildlife to be found in a pinewood includes some rare and unusual species found nowhere else. The mixture of habitats around the woods also adds to the wildlife spectacle, with lochs and rivers among the trees, and adjacent moors and mountains stretching across the landscape. Out of the long list of species which may be encountered I have highlighted a few which are typically associated with pinewoods.

In a native pinewood, the forest floor is usually dominated by lush green carpets of bilberry, or to use its Scottish name, blaeberry

(*Vaccinium myrtillus*), a medicinal plant with edible, sweet, blue-black fruit. Alongside this is heather (*Calluna vulgaris*) which creates a sea of purple when it flowers. Glittering wood moss (*Hylocomium splendens*) fills the shady areas, its yellowy green fronds cloaking the ground, covering boulders and the stems of trees; one of a number of mosses known to have anti-bacterial properties. A characteristic pinewood species is the plume moss (*Ptilium crista-castrensis*), whose long fronds look like the plume in a knight's helmet. Wavy hair grass (*Deschampsia flexuosa*) is a common feature in open areas. In summer its tall slender stems and delicate flower heads can look like a cloud of silver, hovering just above the ground. Creeping ladies tresses (*Goodyera repens*) is a rare orchid, which in the UK is only found in Scottish pinewoods. It has evergreen leaves and in summer bears small white flowers on short stems. Another plant species which is an indicator of ancient woodland is the twinflower (*Linnaea borealis*), mainly found in Deeside and Speyside. This delicate, low, creeping plant has paired, downturned, flowers that look like tiny white wall lamps. This was the favourite plant of the 18th century Swedish botanist Carl Linnaeus, who developed the scientific naming system for plants and animals, and likened this flower to himself, 'lowly, insignificant, disregarded, flowering but for a brief time'.

The pinewoods hold a wealth of insect species and scientists are finding new relationships to other wildlife which are still not fully understood. Blaeberry pollination, for example depends on a rare bumblebee species (*Bombus monticola*). Wood ants are a key part of the insect life of pinewoods, but most notable are their nests. The Scottish wood ant (*Formica aquilonia*) nests consist of large soil domes covered in pine needles, and are up to 1.5 m tall. Within the nest, over 100,000 individual ants make up a colony and their paths can be seen worn into the leaf litter between the trees, as they travel around looking for other insects to eat. The ants are harmless to humans, their main defence being to squirt formic acid across a range of up to 5 cm. Close up, the colony smells like salt and vinegar crisps. It is important not to disturb the nests as these are a threatened species.

The pinewoods are natural refuges for some of Britain's most charismatic mammals, several of which, including red squirrels and

The much loved red squirrel is an endangered species

Scottish wildcat, are now rare and threatened. A number of former inhabitants, such as the elk and brown bear, have long gone.

Red squirrels (*Sciurus vulgaris*) are a typical pinewood species and a large part of their diet is pine cone seeds. It is possible to see them during the day but, failing that, the tell-tale signs of nibbled cones can often be found lying on the ground.

The Scottish wildcat (*Felis silvestris silvestris*), sometimes referred to as the 'Highland Tiger', is an elusive creature whose numbers have been reduced by habitat loss, persecution and cross breeding with domestic cats. Occasionally, the distinctive broad footprints of the wildcat can be seen, especially on snow covered ground. For those who don't mind seeing wildlife in zoos, Scottish wildcats and other pinewood creatures can be viewed at close quarters at the Highland Wildlife Park near Kingussie.

The pine marten (*Martes martes*) is another of the forest hunters. An agile climber, it eats mainly voles and birds as well as occasional squirrels. It is also keen on fruit and honey and one of the best ways of seeing this timid and stealthy predator is when it is lured into the gardens of people living near the forest, with treats such as jam tarts.

There is something comical about watching one of nature's finest hunters bouncing in view to pounce upon an unsuspecting cake.

Red deer (*Cervus elaphus*) and roe deer (*Capreolus capreolus*) are a natural part of the pinewood environment but numbers have been artificially increased over the centuries, not only for sport shooting but also because of a lack of natural predators, such as wolf (Canis lupus) and lynx (*Lynx lynx*). It is not clear when the lynx disappeared but recent evidence suggests they were present in parts of Britain until the 7th century. Wolves remained longer, the last being killed around 350 years ago. Large herds of red deer now roam the hills and glens, descending into the woods for shelter and to feed on the young pines. In many places the density of deer is several times greater than a natural forest can support. Deer, however, are an integral part of a forest's wildlife experience and there is nothing more dramatic than seeing a grand stag perched on a rocky outcrop. The autumn rutting season, when males compete for females, is a noisy affair. The stags' liquid-roar can be heard for miles and the smell of deer musk fills the glens. The smaller roe deer are more solitary and spend most of their time in the forest. Roe deer numbers are also very high through lack of natural predators and they can cause considerable damage to young trees.

Wild boar (*Sus scrofa*), which became extinct in Britain around the end of the 13th century, have been reintroduced in a number of fenced enclosures at pinewood sites such as Guisachan and Alladale. Boar can fulfil an important role in the forest's ecology, breaking up the ground as they dig for food, improving germination conditions for pine and other tree seedlings.

Pinewoods support some of Britain's largest and smallest birds. The golden eagle (*Aquila chrysaetos*) with its wingspan of over two metres can occasionally be seen soaring over the forest and hunting in the more open areas. The forests' other giant bird is the capercaillie (*Tetrao urogallus*), a large grouse the size of a turkey. The males perform dramatic displays to attract females on traditional sites known as leks.

At the other end of the size scale is the crested tit (Parus cristatus), only 8 cm from beak to tail. They feed on small insects and pine cone seeds and nest in the holes formed when branches and stems are torn off in storms.

Britain's only endemic bird (this means it is found nowhere else in the world) is the Scottish crossbill (*Loxia scotica*). A close relative of the common crossbills found across Europe whose bills allow them to extract seeds from various conifer cones, the Scottish crossbill has a strong, deep bill specifically adapted for feeding on tough Scots pine cones.

Early history of the pinewoods

For many people the ancient pinewoods convey the atmosphere of a primeval forest, but scientists and historians have cautioned against viewing these as untouched remnants. Even the most natural of woods in Scotland has had some human influence over thousands of years. Pinewoods are also dynamic, with the woodland boundaries slowly shifting across the landscape in waves of tree death and new growth on the margins. A few, but certainly not all, woods may well have moved several miles from their original location over the centuries. However, with the youngest sites at least several centuries old there is no doubt that, in the words of Steven and Carlisle:

'to stand in them is to feel the past'.

Pine trees can bear seed at the age of 350 years, so it takes less than six generations to reach Roman times. No wonder people feel a sense of awe at the living history in these woods.

The term 'the Great Wood of Caledon' has been referred to in printed texts and maps about Scotland since the 16th century when Hector Boece, then Principal of Aberdeen University, introduced the concept. Many authors, particularly in Victorian times, suggested there had been a single massive forest, often described as rather terrifying, stretching over most of the nation. Historians now largely agree that this is a mythical concept and that there was probably a mixture of open heath, bog and grassland along with broadleaved and conifer woodlands. However, it is clear there were considerable areas of pine forest across the Highlands before a combination of climate change and human action took their toll.

Ancient tree stump preserved in peat and exposed by erosion

Prehistoric Pines

As the last ice age retreated from Britain, the hardy Scots pines together with birch were the pioneering tree species which covered the land. By the time the Romans arrived in Scotland around 2000 years ago, the original forests were much depleted but were still larger than today's. The soil archive shows that Scots pine arrived in southern Britain after the retreat of the last ice age 11,000 years ago. The pine trees first reached Scotland around 9000 years ago and by this time had already spread across much of England, Wales and Ireland but were soon out-competed by oak and alder. By 7800 years ago, pine had largely disappeared from most of England and Southern Scotland.

In the Highlands, Scots pine went through regular periods of expansion and contraction in range due to changes in climate until, around 4500 to 3500 years ago, there was a gradual, widespread decline. The precise causes are unclear but were most likely a combination of climate change, with an increase in wetter weather and expansion of peat formation, along with humans clearing parts of the forest. A common feature in many of the peatbogs that blanket Scotland's hills and glens are the stumps and roots of pine and other trees from this period, preserved for thousands of years by the waterlogged, acidic soils. Stunted pine trees can still grow in bogs as can be seen today at Abernethy and Glen Affric, with hundred year old trees no thicker than an arm, standing a few metres above the peat. Trees and whole woodlands would also have remained on outcrops of better drained ground such as gravel knolls (drumlins) or on steeper slopes where the peat was thin.

The Medieval Forests

After the Romans left, the Norse and Celtic people continued to fell trees for ships and houses and to clear areas of woodland for agriculture. By the end of the medieval period in the 15th century, probably less than 10 % of the original woodland cover remained but the depleted Highland pinewoods were still substantial. Some of the oldest, presently surviving pine trees started as seedlings around that time. Historical evidence suggests that most of the medieval pinewoods still exist today, although their character and structure have been greatly altered by grazing, felling and planting.

There is a surprisingly good historical record of the pinewoods from this period through the works of Timothy Pont who mapped Scotland in the 16th century. Robert Gordon then later revised the work for publication in Blaeu's *Atlas Novus* in 1654. Their maps are accompanied by detailed notes describing not only the pinewoods but also other features and are now available to view on-line through the National Library of Scotland.

The 17th century

The expansion of industrial processes and major wars saw a dramatic and well documented increase in the commercial exploitation of the pinewoods. Making charcoal to fuel iron smelting and glass making furnaces required large amounts of oak, with pine being used as supplies of hardwoods rapidly depleted. The value of the Highland woodland resource came to the notice of King James VI

Loch Arkaig with Glen Mallie in the background

of Scotland and his concerns about over-exploitation led to an Act of the Scottish Parliament in 1609, preventing the use of wood for iron making. The Parliament wanted the forests better managed to provide timber particularly for ship building at a time of great sea battles.

Survivors of the Jacobite Uprising

The period of Britain's last civil war in the mid 18th century, when Charles Edward Stuart, 'Bonnie Prince Charlie', attempted to regain the throne of England and Scotland for the House of Stuart, features strongly in the history of the pinewoods. Many of the pine trees surviving today were alive during those troubled times. Remote wooded glens in the west Highlands provided hiding places and escape routes for the Prince and his supporters. Associated with many of the pinewoods, such as those at Glen Moriston, Glen Strathfarrar and Loch Arkaig are great tales and myths from that time, which are now an important part of the Scottish tourist experience.

The Napoleonic wars in the 19th century saw more timber removed from many of the pinewoods after higher custom payments made Scandinavian imports more expensive. Most of the woods in the east regenerated as grazing animals were largely excluded from the forests. Interestingly, some of today's surviving pinewoods have regular, evenly aged stands of trees from this period. The west Highlands however, saw greater losses with many of the trees unable to regenerate as the land was favoured for livestock grazing.

Further heavy felling took place during the First World War when domestic timber was required for shoring trenches and to replace the supplies of foreign timber which were being blockaded. With military precision, companies of foresters would clear thousands of trees in a few months, leaving whole hillsides bare. A few decades later, the Second World War also saw vast areas of woodland cleared to provide timber for the war effort. Sometimes, damage occurred as a result of unfortunate incidents where army training exercises caused fires. The dead remains of hundreds of trees from one such fire at Glen Mallie can still be seen.

At the end of the First World War a new era of state controlled forestry began with the establishment of the Forestry Commission in 1919, aimed at replenishing the country's diminished timber resources. Forest land, including some of the old pinewood sites, was acquired for the Government by the Forestry Commission. The management aims at these sites had little to do with retaining the character or ecology of the woods, but were focussed on providing straight, tall trees as a strategic reserve in case of another war. The pinewoods suffered a further blow when the foresters at that time decided to plant conifer species introduced from North America. These 'alien' or non-native conifers are not part of the natural ecology of Scotland and do not support the same wildlife as the pine trees.

The enemy within

The management of livestock and deer has had, and continues to have, a major influence on the wellbeing of the pinewoods. For thousands of years the pinewoods were grazed by domesticated cattle, sheep

Red deer stags

and goats and, as the Highland population expanded to a peak in the 19th century the depredation, particularly from cattle in unprotected forests, grew steadily worse. There was even greater pressure when sheep flocks became much larger and new breeds were introduced. This was the time of the Clearances when many Highland families were forced 'or felt compelled' to leave their ancestral homes to make way for intensive sheep farming.

In the late 19th century the pinewoods' value as a timber resource crashed as cheaper Scandinavian imports became available. Sheep farming also became less profitable at this time and Scotland's Victorian landlords, looking for new sources of income, began to manage the pinewoods to provide sport shooting, or deer stalking. The deer herds as well as having few natural predators were deliberately increased in numbers by management, such as the provision of winter feed.

The result of the deer management was that whilst mature trees remained standing for hundreds of years, a subtle but deadly change was taking place, with few young trees surviving such intensive grazing.

Over the decades, the structure of the wood changed from that of a thriving family of different ages to one where all that remained were the old grandparents. When trees were felled or destroyed by fire, the natural flush of young growth would be halted in its early stages by grazing to such an extent that no new woodland could establish a hold. This intensive grazing amounted to deforestation by stealth, a problem which remains in some parts today.

Pinewoods from the middle 20th century to the present day

After the end of the Second World War, the devastated condition of the pinewoods caught the attention of foresters and conservationists. Initial efforts to address the problem focussed on protecting the trees as a timber resource but, as UK and international wildlife conservation laws developed, the woods were also recognised as a valuable habitat.

In 1950 Professor Steven and Jock Carlisle embarked on a six year study of the pinewoods to provide a full account of their location, condition and history. Their book *The Native Pinewoods of Scotland* published in 1959, highlighted the perilous state of the pinewoods and the authors argued that unless urgent action was taken to control the impact of sheep and deer grazing on the young trees, the pinewoods could be lost to future generations.

The 1960s and 70s saw a massive change in the Scottish landscape, with thousands of hectares of moorland in and around the pinewoods being ploughed and planted with Sitka Spruce (*Picea sitchensis*) and other non-native conifers. Aided by the Forestry Commission, private and state owned land was turned over to swathes of evenly aged trees. The ancient practice of forestry management was being replaced by intensive tree cropping.

In the 1970s conservationists, ecologists and foresters recognised that something more had to be done to halt the decline of the native pinewoods and expand their area. Government grants were awarded to landowners to restore the native pinewoods and increase their area. To address the deterioration of the woods from the ever-present impact of grazing animals, grants were made available for conservation measures such as the erection of fencing to exclude deer and sheep. These fenced areas, known as 'exclosures', are intended to protect the young pine trees by keeping grazing animals out, as opposed to 'enclosures' which keep animals in.

In 1987 I was employed by the Royal Society for the Protection of Birds to re-examine the sites studied by Steven and Carlisle thirty years earlier. My final report for all 35 sites showed that over a quarter of the original woods had been damaged through ploughing and planting with non-native conifers. Large areas of forest had been felled for timber, or for the development of hydroelectric generation. Having once dominated parts of the Highland landscape, only 12,000 ha of ancient pinewood remained; less than 0.2 % of the area of Scotland. Most of the woods were small, less than 100 ha, and some consisted of only a few dozen trees. Unfortunately, the conservation grants introduced under the Native Pinewood Grant Scheme nine years earlier had not been widely taken up by landowners.

The 1990s saw an end to major pinewood losses from felling and planting with non-native conifers. Over 80% of the ancient pinewood area was within protected sites and several private owners had agreed to encourage natural regeneration.

In 1998 the Forestry Commission published The Caledonian Pinewood Inventory based on extensive field work by Graham Tuley, which identified 84 pinewood sites, covering a total of 18,000 ha. The new survey included large areas of open ground suitable for tree regeneration, among scattered mature trees. It also increased the list of sites by subdividing some woods and adding others which Steven and Carlisle had excluded because their origins could not be confirmed.

The arrival of the 21st century provided an opportunity to mark a new future for Scotland's native pinewoods. The 'Millennium Forest' for Scotland initiative was launched to fund native woodland projects ranging from major restoration works in estates such as Glen Affric to the planting of small community woodlands.

In 2003, a new Scottish Forestry Grant Scheme was introduced with a focus on the expansion of the existing woods by natural regeneration. Within the Forestry Commission Scotland sites a highly visible conservation effort was undertaken on a massive scale. Non-native plantations in the old pinewoods were felled and left to waste, the small trees being of little commercial use. The resulting scene was not pretty. At sites such as Glenmore, Glen Garry and Glen Einig, bleached stems and branches on the ground formed an impenetrable carpet. This effect is temporary and in time the material will rot and woodland plants will return with young pines regenerating from nearby seed sources.

One of the most heart-warming features of the conservation work is the convincing evidence that control of grazing animals leads to spectacular regeneration of the pine and its broadleaved companion species. Even in the extreme wet, west coast, conditions of Barisdale and Beinn Eighe there is rich growth of young trees within the fenced deer exclosures. With deer numbers having doubled in Scotland in the last 50 years, fencing is seen as a short term measure. Fences are fallible, and large birds such as capercaillie and black grouse can be killed after colliding with the wires. Deer Management Groups have been established as part of a strategy to control deer numbers so that the woodlands can thrive.

Climate Change

The start of the 21st century saw a growing awareness of climate change as an overwhelming environmental threat. For the pinewoods, which are adapted to grow in a cooler climate, a warmer future may hold serious threats. Hotter summers mean greater risk of fires and temperature change may benefit insect and fungal pest species, such as needle blight, which infest the trees and can kill them.

A perverse threat however is the damaging response of those who would prematurely abandon conservation effort in the pinewoods, assuming them to be doomed by climate change. There is considerable uncertainty about the precise impact of climate change and it is too soon to be giving up on a species which has survived here for over

9000 years. The fact that individuals can live up to 500 years also means that, although stressed, some pine trees may survive the worst of the changing climate. We should therefore seek to make our few remaining natural habitats as robust and healthy as possible. In a damaged state, the woods and their wildlife will certainly be more at risk of extinction.

Looking to the future

The fifty years since 1960 have seen a complete turnaround in the fortune of the pinewoods. Their conservation importance has been established in law and they are no longer under threat from large scale felling. Much of the habitat is being restored and expanded. It has taken some time but at last the call of Steven and Carlisle 'to take positive action to protect and regenerate these woodlands' has been heeded.

A Forestry Commission Scotland review of native woodland restoration on its estate in 2004 set some pointers for the future, including restoring a more diverse mix of habitats and age structure to existing woods, expanding native woodland habitats at a landscape scale, planning for sustainable timber production and monitoring conservation progress.

A big challenge for pinewood conservation is in managing them not just as trees but as whole ecosystems, with their typical species and habitats. Areas of open ground such as peatbogs and heather moor are a natural part of the forest, and work is already underway in a few pinewoods to rewet damaged peatland by blocking old drainage ditches. Allowing the woods to shift naturally between dominance of different tree species is also an important consideration. There is evidence of some sites where pine and birch communities periodically replace each other over the centuries. Past management has resulted in the loss of the natural broadleaved tree species, such as oak, rowan and juniper which may need to be replanted to restore a more natural species mix.

Large forest areas extending up to the natural tree limit in the mountains are a rare feature in the UK. Expansion of the woods from Strathspey to Deeside is enabling trees to recover lost ground around

Mature pinewood habitat, The Black Wood of Rannoch

the fringes of the Cairngorms and extend up the slopes to 600 m or more. Covering several thousand hectares and composed of open ground, broadleaved trees and pinewoods this is a prime example of landscape scale conservation and the result is one of the most popular, wildlife rich and thriving areas in the Scottish Highlands.

Public access to the pinewoods requires careful management to allow visitors to experience these ancient natural wonders whilst also avoiding disturbance to wildlife and damage to habitat. There is a risk of fires which can have long lasting, devastating effects and the problem of litter, notably visible at Loch Arkaig where overnight visitors too often leave piles of discarded fishing gear, bottles, cans and plastic. Such problems can be addressed however, and there is a long history of successfully managing access within important wildlife areas. At the Black Wood of Rannoch, Forestry Commission Scotland provide for the casual day tripper, the keen walker and cyclist, and still offer a wilderness experience for brave-hearted wild campers.

Zoning of activities within pinewoods to provide strict conservation areas, recreation areas, or commercial wood production is an important part of planning for their future, although it is not necessary to achieve all these goals in every part of the pinewood, or in every pinewood. It is right that some woods should be left unharvested to provide a natural benchmark. For some, the idea of sustainably managed forestry needs to be complemented by more of a wilderness approach. The Trees for Life charity has as its vision the establishment of a wild wood in the west Highlands around Glen Moriston; a wood managed for its own sake as a home for wildlife with no commercial tree harvesting activity. With so much of our countryside managed for commercial timber or sporting interests why not have a few places where the natural environment has primacy?

As conservation effort continues, monitoring progress and assessing the extent and condition of the pinewoods is essential. A new comprehensive Native Woodlands Survey of Scotland has been published by the Forestry Commission Scotland and supported by Scottish Natural Heritage. Based on extensive field surveys, this is one of the largest habitat assessments ever undertaken in this country and provides detailed maps showing locations of the woods.

Visiting the woods

The pinewoods offer an exhilarating experience whatever the time of year. In spring and summer the scent of pine combines with the sound of songbirds in the trees and buzzards 'mewing' overhead. Purple flowering heather in autumn surrounds the woods which are a colourful tapestry of green pine contrasting with the reds and yellows of alder, birch and rowan. In winter the pinewood becomes a Christmas card snow scene.

The survival of these ancient pinewood remnants has in part been due to their remote locations often in isolated and mountainous terrain, so care is needed in visiting them, particularly as the weather in the Scottish Highlands can be extreme and changeable. It is good practice when travelling in remote areas to set up a 'buddy system' where a description of the journey plan is given to a friend with an arranged time to make phone contact.

There is an intense feeling of achievement to be derived from getting to a site under your own effort. The statutory right of access in Scotland means walkers, cyclists, and horse riders are permitted on most land provided they act responsibly. This includes having consideration for the management of estates, particularly during sensitive times such as the deer stalking season or during forestry operations. *The Scottish Outdoor Access Code* explains these rights in detail and emphasises the need to respect the environment. The woods and moors are sensitive habitats where wildlife can be disturbed to the point of failing to breed that year, and rare plants can be damaged. Several of the pinewoods have been burned to the ground as a result of careless use of campfires.

Most of the larger pinewoods now cater for a wide range of visitors with easy access routes and excellent facilities. Forestry Commission Scotland has major centres at Glenmore and Glen Affric receiving over 100,000 visitors a year. There are also the showcase National Nature Reserves such as Beinn Eighe and the RSPB's Abernethy Forest along with the privately owned Rothiemurchus Forest and Glen Tanar providing a resource for local people and tourists to enjoy and bringing considerable economic benefits.

The ancient pinewoods can be reached using established paths, many of which are centuries old. These routes, now identified as Heritage paths, include the drove roads which were used from the 16th to the 19th century to walk cattle from their grazing lands to the markets or 'trysts'. In remote Highland communities coffin roads allowed people to carry the boxed bodies of corpses for burial in consecrated ground, which was often many miles away. Before the middle 18th century most of Scotland's main roads in the Highlands were nothing more than cattle tracks. From 1725, infrastructure improved as General Wade, Commander in Chief of the army in North Britain, began a campaign of road and bridge building to allow the movement of his troops. Some of the original features of these old roads such as stone bridges

Abernethy Forest looking south to Cairngorm, Strath Nethy and Bynack Mor

and culverts can still be seen on the old military roads such as the one leading from Fort Augustus to Glen Moriston.

One of the best ways to reach the farthest flung pinewoods is by cycling using routes away from traffic on major roads. Sustrans has identified a National Cycle Network in Scotland of quiet public roads as well as establishing dedicated off road cycle routes. The Sustrans website also gives details of the other major long distance routes such as the Great Glen Way. Coaches and buses also provide an excellent service for reaching most of the towns and villages across Scotland. Some will also carry bikes, usually with prior booking.

Despite the closure of many train lines in the 1960s, Scotland still has a rail network that reaches to the extremities of the Highlands. It is no exaggeration to say that one of the world's most spectacular train journeys runs on the West Highland Line from Glasgow to Mallaig.

West Highland Line, Glenfinnan viaduct

This great example of Victorian engineering includes the Glenfinnan viaduct (of Harry Potter fame), passes castles and lochs and skirts round some of Scotland's highest mountains. Dotted along the route are several of the ancient pinewoods, with Glen Falloch and Tyndrum visible in the distance. The train passes right through one of the Black Mount woods at Crannach and in the days of steam trains, red hot cinders from the funnel often started fires serious enough to kill many trees. The railway line from Perth to Inverness conveniently passes by the main Strathspey pinewoods, Aviemore, Glenmore and Rothiemurchus, which can be seen clinging to the slopes of the Cairngorms.

From Inverness, the famous Kyle of Lochalsh line heads west on a lonely single track across wild, boggy moorland, to the tiny station beside the wood at Achnashellach. Ironically it was this line which allowed vast amounts of timber to be removed from the pinewoods around Loch Maree and Achnashellach during the Second World War. In a short guide to the Kyle line, the climber and writer Tom Weir wrote in 1971 that these scenic rail routes should be promoted as tourist attractions. Trains provide an excellent window onto our natural history. After a hard day's hiking in the cold, it is intensely comforting to see the carriages approaching out of the hills with the promise of warmth and relaxation. For the cyclist, train companies advise that it is important to book in advance, as spaces are limited, but most trains do offer secure facilities for bikes.

Travelling without a car means that long trips will require overnight accommodation. There is a variety of low cost options available, even in the most rural areas. Some of the best opportunities to meet interesting people are in the campsites, hostels and guest houses. At the more upmarket end there are the old Inns, some dating back to the 18th century, and ornate stone Victorian sporting hotels. Open fires, food, good local beer and a huge selection of whiskies can be had in many of these old hostelries which have long catered for the weary, hungry traveller.

One of the less appealing features of a visit to the pinewoods, as with any outdoor location in the Highlands, are the dreaded midges. They are most active in the early morning and at dusk with the worst biting time in the Highlands being from June to August. The level of midges varies depending on weather and location but there is a Scottish midge forecast available in newspapers and online. The other pest is the tick, a small blood-sucking invertebrate which can be found in long vegetation, particularly heather and bracken. Insect repellents, thick socks and gaiters help deter them. Scottish Natural Heritage provides further information on how to deal with ticks and reduce the risk of catching Lyme's disease from the tick's bite.

The ancient pinewood sites included in this book are those that were identified in the 1950s by Steven and Carlisle as truly native. For me these hold a special attraction as there is little doubt they are descendants of the original forests with their histories reaching back thousands of years. There are other small pockets of ancient pine woodland dotted around Scotland, although some may have been planted only a few centuries ago. These woodlands are still fascinating to visit and are an important part of Scotland's rich natural heritage.

Using this guide

The pinewoods are divided into seven groups consistent with those suggested by Steven and Carlisle, as these are a convenient way to present sites with similar access routes. Modern science has identified different groups based on the biochemical characteristics of the trees but these are less useful for a guide book.

The start of each group has a regional map locating the woods and a brief summary of the area. The individual woods have maps showing the main topographical features and suggested routes are given. These maps are intended to give a basic idea of the geography, to help with planning a visit and should not be used for navigation. The relevant Ordnance Survey (OS) map sheet is given at the start of each site account. Further details on safety, access, travel and accommodation are given at the end the book.

Storm and flood damage, particularly in winter, can cause disruption with closure of paths and bridges. Always check that routes are available before heading out on a journey.

Key to maps

▪ Ancient pinewood area ('Core Area' including scattered trees, from Caledonian Pinewood Inventory 1998).

▪ Other woodland

– – Walking and cycling routes given in the travel notes.

– – Suggested connecting routes between pinewoods.

NB these maps are provided simply as a guide to planning routes and should not be used for navigation.

Regeneration exclosure with fence to keep deer out, Strath Vaich

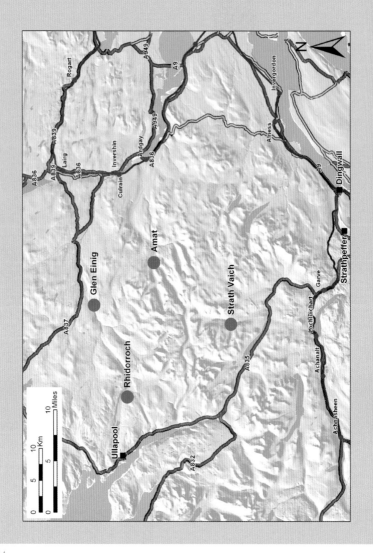

NORTHERN GROUP

For thousands of years pinewoods, along with oak and birch, grew across much of Sutherland but climate change and human activity reduced their northern extent leaving only a few pine trees around Ben Hope. Elsewhere, small remnants lie secluded among remote hills and glens. There is a charm to these isolated pinewoods, so far away from major tourist facilities. Although the northern group are among the smallest of all the ancient pinewoods, recent conservation effort has resulted in a welcome regeneration of young trees. In the public and privately owned sites, there are ambitious plans to extend the pinewoods in a return to their former glory.

Ancient routes through the glens provide access for the experienced hillwalker and cyclist to be rewarded by 'oases' of old pines set perfectly by nature in a vast moorland landscape.

Small groups of pine trees can also be found along the east coast at the Scottish Wildlife Trust, Loch Fleet reserve and the Woodland Trust reserve at Loch Migdale. Although these may be survivors of 18th century planting they still have natural characteristics and are well worth visiting.

- Buses from Inverness to Ullapool provide good connections and some take bikes if booked in advance (www.decoaches.co.uk, www.travelinescotland.com).
- The nearest railway stations are in the east at Ardgay and Invershin (see Scotrail).
- Accommodation: there are several old inns and guest houses (see VisitScotland).
- **Pinewoods with major visitor facilities:** None

RHIDORROCH

Map: OS 1:50,000 Sheet No 20
Wood Grid ref: NH 235 933
Access Point: Ullapool (NH 128 949)

The pinewood lies in the tranquil Glen Achall near the purpose built 18th century fishing town of Ullapool. Once the gathering ground for driven cattle heading to the markets in the east the glen holds many delights from broad flood plans to steep crags and majestic waterfalls. Birch and pinewoods lie to the east of Loch Achall.

A remote mountainous area in the west of Scotland which is home to golden eagle, otter and pine marten. The mouth of this quiet and gentle glen with its wide floodplains soon gives way to the dramatic grandeur of steep mountain slopes. A good track follows Rhidorroch River to the main pinewoods around East Rhidorroch Lodge. Ancient birchwoods grow alongside the pine to form a lovely woodland backdrop. Pine trees also cling to the steep scree slopes of Creagan Ghiuthas which is Gaelic for 'crag of the pines'. A narrow suspended cable bridge crosses the river and a rough steep path continues through some mature pines up to the Easna Baintighearna (waterfall of the lady).

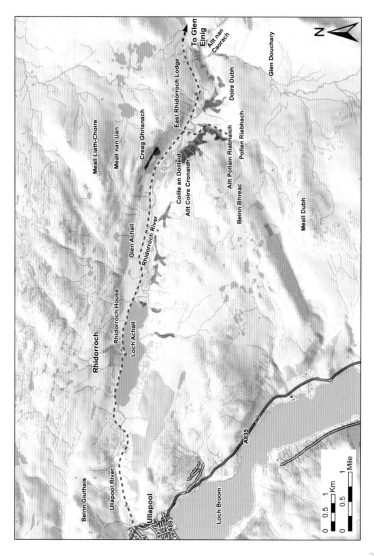

Travel Notes

The nearest town is Ullapool which can be reached by bus from Inverness. The service carries bikes, either in the back of the coach in the quiet season or on a trailer in the summer months. The journey takes an hour and a half and ends at the Ullapool ferry terminal on Loch Broom. The wood is reached by taking the estate track to Rhidorroch Lodge, entered by turning right off the A835 (NH 128 949) approximately 1 km north of Ullapool. This is a well used route by walkers and cyclists travelling between Ullapool and Strath Oykel in the east.

The track to Rhidorroch starts with a disconcerting limestone quarry operation with heavy vehicles using this first part of the route. Once clear of the quarry the track soon reaches the calm flat grassy shores of Loch Achall and a bridge over the Ullapool River. After passing the loch, the route continues east below the cliffs of Creag Ghrianach and alongside the Rhidorroch River. Across the river is the birchwood at Coill' an Doilleid which contains a scattering of mature pine trees. At the end of the wood lies East Rhidorroch Lodge, reached by a suspended cable bridge (NH 236 936) over the river. A path goes up the Pollan Riabhach to a dense area of mature pines and a spectacular 80 foot (25 m) waterfall, (Easna Baintighearna). Retrace back down over the bridge to the main track and continue east, along the north shore of the river. The track becomes very rough with a relentless rise up the steep slope towards Loch an Daimh. After a few kilometres the track levels out and the birch wood with scattered pines at Doire Dubh can be seen. More individual, mature pine trees lie scattered within Glen Douchary and Allt nan Caorach, which is where the Rhidorroch pinewood ends.

There are guest houses in plenty in Ullapool. The Carnoch guest house, on West Argylle Street is a renowned, charming traditional cottage. For those on a more generous budget, Rhidorroch estate offers self-catering cottages, and accommodation for large groups is available in the Victorian, Rhidorroch House.

A route from Rhidorroch to Glen Einig

From the north side of the Rhidorroch River in front of the old Lodge (NH 236 936) head east along the track to Loch an Daimh. The going is fairly rough and stony, with wet boggy patches and several small fords which are, usually, easy to cross. Blanket bog and heather moor extend all around this isolated landscape, with only a few patches of birch trees including several dead individuals dotted along the hillside across the loch. Eventually the two chimneys of the Knockdamph bothy come into view. The track continues alongside the Abhainn Poiblidh with a ford crossing (NH 320 978) and then heads east to the Duag Bridge at the foot of Strath Mulzie. After crossing, take the left fork along the vehicle track into the forest at the west end of Glen Einig pinewood.

Gaelic Place Names

Rhidorroch	the dark shieling	Easna Baintighearna	waterfall of the lady
Strath Mulzie	mill-strath	Pollan Riabhach	the speckled little pool
Abhainn Poiblidh	river of the booth	Coill' an Doilleid	wood of darkness
Loch an Daimh	loch of the stag	Creag Ghrianach	the sunny crag
Innis Dhonacal	Duncan's meadow	Meall nan Uan	hill of lambs
Allt nan Caorach	burn of the sheep		

Pine and birchwoods in Glen Achall

GLEN EINIG

Map: 1:50,000 Sheet No 20
Wood Grid ref: NH 365 988
Access Point: Oykel Bridge (NC 385 008)

The most northerly remaining ancient pinewood in Scotland with scattered trees along the banks of the River Einig. This is part of a Forestry Commission site undergoing conservation management with good paths in a remote heather moorland setting.

Glen Einig in the far reaches of the Highlands is a small yet inspiring woodland landscape in the process of transformation. A few hundred mature pine lie scattered across an area that was once heavily planted with commercial non-native conifers. These 'aliens' have now been removed to allow the pines and broadleaved trees to regenerate as part of a major native woodland restoration plan for the area. There are walks along the south bank of the River Einig site to Duag Bridge which offers excellent views back over the glen. The old stone built Oykel Bridge, a few yards from the current modern one which now carries the A837 is well worth a visit. An old school with its corrugated roof and abandoned crofts dotted around the moors are a reminder of the fact that these isolated areas once held numerous families.

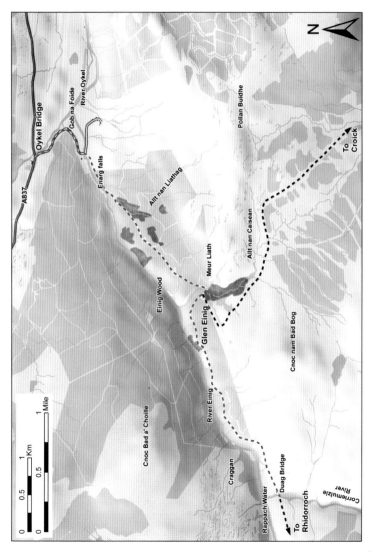

Travel Notes

The nearest train station is at Invershin, and from here it is 20 km by road on the A837 to Oykel Bridge (NC 385 008). After crossing the bridge, there is a surfaced road to the left, which heads south on the west bank of the River Oykel for just under a kilometre before ending at the cottages at Gob na Foide. Here the route becomes a good forest track which turns left to cross over the River Einig and then follows the south shore into the Forestry Commission Scotland woodland. The first group of mature pines can be seen where the Allt nan Liathag flows under the track about half a kilometre into the wood. The track dips down after a few more kilometres into the Allt nan Caisean where a large area of mature birch and pine can be seen extending out beyond the deer fence at Meur Liath (NH 336 983). After crossing the stream here it is worth taking the track that turns south west through the fence, then following it south for a few hundred metres to see the main group of old trees. Here you also gain a fantastic panoramic view of Glen Einig with the plantation stretching into the distance. Go back down into the forest and continue west alongside the River Einig to the far end of the wood at Duag Bridge. The return journey to Oykel Bridge is back along the same route.

Glen Einig native woodland regeneration

Oykel Bridge has an old Victorian hotel, popular with anglers. There are two bridges: the present modern one, over which the A837 travels and just beside it is the old 19th century stone single arch bridge which is well worth a look. The Achness Hotel, 10 km east, along the A837 has accommodation in the old stable blocks. There are also a number of guest houses in Rosehall.

A route from Glen Einig to Amat Woods

From Oykel Bridge go through Glen Einig forest to the Allt nan Caisean and then exit the forest where the track heads south east towards Strath Cuileannach. This vehicle track has a long steep rise for about 2 km then enters a fenced forestry plantation where the going is rough for another 2 km before leaving the plantation. The track then follows the north shore of the Black Water and passes some very openly spaced young Scots pine plantations. At Lubachoinnich farmhouse (NH 414 954) the track becomes easier and continues for another 6 km to Croick. From Croick the road goes south east for 2 km to a junction and a public telephone box. Take the turning to the south over a bridge crossing the Black Water and continue to Amat forest.

Gaelic Place Names

Glen Einig	young bird glen	Meur Liath	the grey rivulet
Lubachoinnich	the mossy loop	Allt nan Liathag	burn of the salmon-trout
Strath Cuileannach	the strath of the litter		
Allt nan Caisean	burn of the cheeses	Corriemulzie	mill corrie

AMAT WOODS

Map: OS 1:50,000 Sheet No 20
Wood Grid ref: NH 468 902
Access point: Craigs (NH 473 909)

A large series of woodlands between Amat and Glen Alladale at the head of Strath Carron include some of the most majestic and finest stands of ancient pinewood in the north of Scotland. This is a wonderful example of thriving ancient pinewood habitat in a spectacular mountain landscape.

Good surfaced roads and tracks lead from Strath Carron with its grassy croftland to the dramatic mountains and moorland with large old pines alongside the rapid waters of the River Carron. Here amid the tumbling white waters there are waterfalls and deep pools where salmon can be seen leaping. The scene then shifts to the gentle meandering River Alladale with a pinewood standing like a lone sentinel guarding the dark head of the glen. Alladale estate is home to a unique forest experiment aimed at introducing animals such as boar and elk which were once part of the ancient forests special wildlife. Strath Carron has several places offering accommodation. The small church at Croick is worth a visit to learn about one of the sad tales from the era of the Highland Clearances.

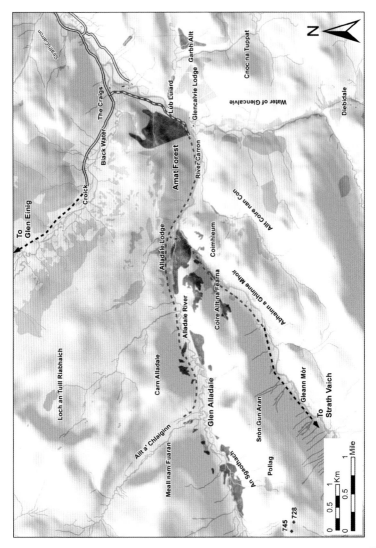

Travel Notes

The woods lie some 12 km west of the railway station at Ardgay and can be reached along a surfaced road running north-west from the A836. Cross the bridge at Cornhill and turn left along the north shore of the River Carron. There is a junction at the Craigs (NH 473 909) with a small church building and a telephone box. Here the route crosses south over the bridge with the Amat forest visible on the right. Continue along this road west past Glencalvie Lodge. There is a small path to the left which is worth a detour down to the Eas Charron waterfall with pine trees lining the gorge. Returning to the road, continue west to the entrance of the Alladale estate. The track skirts below Alladale Lodge and heads west. At the point where the Allt Riabhach joins the Alladale River there are couple of cottages and a new vehicle track which is followed for 3 km to the confluence of the Alladale River and the Allt a Chlaiginn (NH 406 890). The An Sgaothach pinewood can be seen to the south west in the shadow of the steep mountain slopes below Lochan Pollaig.

An Sgaothach at the head of Glen Alladale

Returning eastwards back to Alladale Lodge the experimental fenced enclosure to hold elk and other animals can be seen on the south side of the river. Just past the lodge there is a track to the right that crosses the river and heads south west alongside the Abhainn a Ghlinne Mhoir to give a view of those scattered trees on the east bank, which survived a major fire in the 1940s.

Strath Vaich pinewood, which lies to the south, can be reached by following this heritage path, an ancient road through Gleann Mor past Deanich Lodge and down the east side of Loch Vaich.

There is guest house accommodation around Ardgay. Along Strath Carron are guest houses including Corvost, an old croft which used to be a common feature along the valley but now most are gone.

Gaelic Place Names

Lochan Pollaig	loch of the little pool	Eas Charron	falls of Carron
An Sgaothach	the swarm	Gleann Mor	the big glen
Allt a Chlaiginn	burn of the skull	Abhainne a Ghlinne Mhoire	river of the big glen
Allt Riabhach	the speckled burn		

STRATH VAICH

Map: OS 1:50,000 Sheet No 20
Wood Grid ref: NH 345 745
Access Point: Black Bridge carpark (NH 373 708)

Hugging the west shore of Loch Vaich this small pinewood holds some splendid mature pines and is undergoing restoration management in parts, to protect young trees. This remote and high mountainous area is home to golden eagle and ptarmigan.

From the main Inverness – Ullapool road a good surfaced route leads up through this quiet glen. On the low ground, farmland with herds of highland cattle grazing alongside the river gives way to steep moorland and crags. The track eventually reaches a stone dam at the end of Loch Vaich which was developed in the 1950s to provide hydro electricity generation. In the peat around the loch, erosion has exposed the preserved roots and trunks of pine that grew here over 4000 years ago. A hill track leads up the east side of the loch on an ancient route to Glen Alladale and provides excellent views of the pinewood.

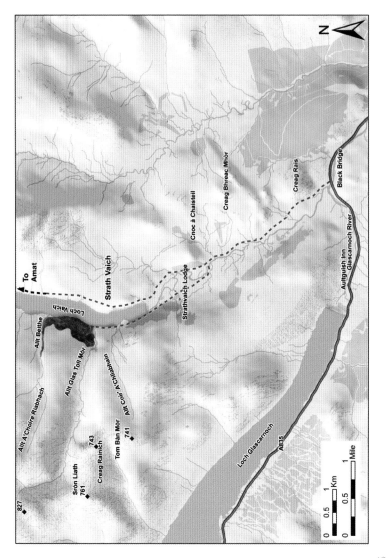

Loch Vaich with c4000 year old pine stump

Travel Notes

The nearest railway station is at Garve. Take the A835 north past the Inchbae Lodge hotel to the Black Bridge car park (NH 373 708). An estate road leads north-west up to Loch Vaich, past Strathvaich Lodge and ends at the dam on the south shore of the loch. The wood is situated between Allt Glas Toll Mor and Allt Glas Toll Beag, on the west of the loch, and can be reached by a narrow path which is rough and wet and best done on foot. Alternatively the wood can be viewed from the east shore of the loch where the track is more passable by bike. Return by the same route back to the Black Bridge car park.

There are a number of famous old public houses, including the Aultguish Inn along the A835 which is the main route between Inverness and Ullapool and the northwest.

Gaelic Place Names

Strath Vaich	strath of the byre	Allt Glas Toll Beag,	small burn of the grey hole
Aultguish	the fir burn	Allt Coir' a Chliabhain	burn of the corrie of the creel
Inchbae	the birch island		
Allt Glas Toll Mor	big burn of the grey hole		

Scottish wood ant nest

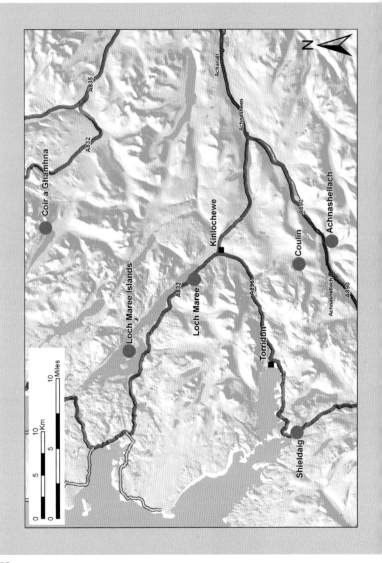

WESTERN GROUP

The western pinewoods lie in a spectacular landscape among the wild and remote Torridon mountains and are also found at sea-level near Shieldaig at the western limit of Scots pine's northern European distribution.

Although these parts are far flung, Queen Victoria and writers including James Hogg, the Ettrick Shepherd, made their way here in the 19th century and marvelled at the pinewoods. Wildlife conservation in its legal form began with the declaration of Britain's first National Nature Reserve at Beinn Eighe.

Lesser known but no less appealing is the Coulin pinewood with its beautiful reed fringed lochs and mountain backdrop. There is also a 'hidden' pinewood, Coir' a' Ghamhna, tucked away in a steep corrie below the towering pinnacles of An Teallach. A strenuous but deeply satisfying walk leads to a wonderful spectacle of pine trees clinging to the very foundations of this magnificent mountain.

The single track Kyle railway line from Inverness offers access to most of these far outlying woods and is an interesting experience in itself as the route crosses open expanses of desolate moor passing small groups of old pine trees such as those at Achanalt.

Achnashellach pinewood is conveniently close by the railway station of the same name. There are also buses from Inverness to the route starting points at Kinlochewe and Dundonnell.

- Accommodation: Hotels and guest houses around Torridon, Kinlochewe and Shieldaig (see VisitScotland).
- **Pinewoods with major visitor facilities:**
    Beinn Eighe: Scottish Natural Heritage.
    Also National Trust for Scotland, Torridon Countryside Centre.

SHIELDAIG

Map: OS 1:50,000 Sheet No 24
Wood Grid ref: NG 820 524
Access Point: Shieldaig, Loch Torridon (NG 815 535)

The pinewood clinging to the steep rocky slopes of Ben Shieldaig is known as Coille Creag Loch and is the most westerly of all Scotland's ancient pinewoods. The trees extend down to the main road where several magnificent specimens over 250 years old line the shore of Loch Dughaill.

The village of Shieldaig, tucked into the shore of Loch Torridon, was purpose built in the 19th century to train soldiers for the Napoleonic wars. Now it is a picturesque and peaceful retreat with excellent shoreline walks. There are also boat trips available around Loch Shieldaig to see the dramatic scenery and amazing wildlife including white-tailed sea eagles. Shieldaig Island with its impressive stand of tall Scots pine is thought to have been planted to provide poles for drying fishing nets.

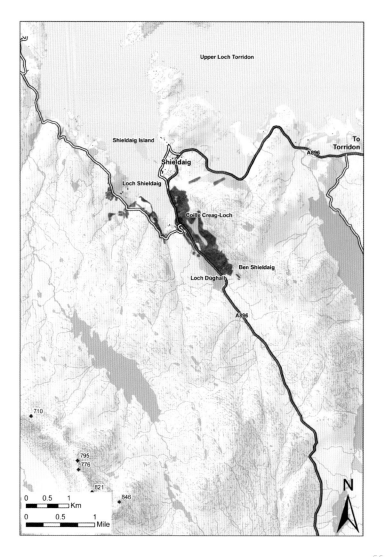

Coille Creag Loch on the slopes of Ben Shieldaig

Travel Notes

The nearest railway station is Strathcarron on the Kyle line between Inverness and Kyle of Lochalsh. From this small, remote station it takes a fairly strenuous 20 km cycle along the A896 to reach Shieldaig. The route along Loch Carron is more level before a long climb out of Lochcarron village. Once over the summit there are spectacular views over Loch Kishorn. A few kilometres further it is worth stopping at the Rassal Ash wood (NG 840 431). This is one of the most northerly examples of an ancient, native ash wood. From here, another slow ascent along the A896 takes the cyclist to the head of Glen Shieldaig where the pinewood can be seen lining the steep slopes of Ben Shieldaig. The road continues on past the main wood at Coille Creag Loch along the shores of Loch Dughaill to Shieldaig.

There are guest houses and Inns in Shieldaig and 10 km along the road at Torridon and Annat. The Torridon Hotel, once a grand shooting lodge, has a public bar and food.

A route from Shieldaig to Loch Maree

From Shieldaig continue 25 km, largely uphill, along the A896 to Kinlochewe. The route goes through the dramatic Glen Torridon with its towering, rocky sides rising to the 1000 m summit of Liathach. A short distance further there is a spectacular viewpoint looking over to the equally impressive Beinn Eighe. The road continues past Loch Clair, fringed by pinewoods, at the entrance to the Coulin estate and then skirts the southern slopes of Beinn Eighe before reaching the junction with the A832 at Kinlochewe. Turn left along this road to the Beinn Eighe Visitor Centre (NH 019 630) at the head of Loch Maree.

Gaelic Place Names

Liathach	grey one	Loch Dughaill	Dhughall's Loch
Coille Creag-loch	crag loch wood		

LOCH MAREE

Map: OS 1:50,000 Sheet No 19
Wood Grid ref: NH 010 609 to NG 897 735
Access Point: Beinn Eighe Visitor Centre (NH 019 630)

On the shore of Loch Maree sits a large pinewood, Coille na Glas Leitire, which was the first National Nature Reserve to be designated in Britain. Large islands in the loch also support magnificent pine trees.

The spectacular Torridon Mountains convey a true wilderness experience. The distinctive summit of Beinn Eighe with its shimmering white quartzite slopes contrasts dramatically with the bottle green canopy of the mature pinewood. Loch Maree is a large freshwater loch whose islands have an enchanting history encompassing religious hermits, a Viking love tragedy and a pagan wishing tree. The Scottish Natural Heritage visitor centre at Beinn Eighe near Kinlochewe provides an excellent interactive experience on the wildlife and history of the area. There is a variety of marked trails to suit all abilities.

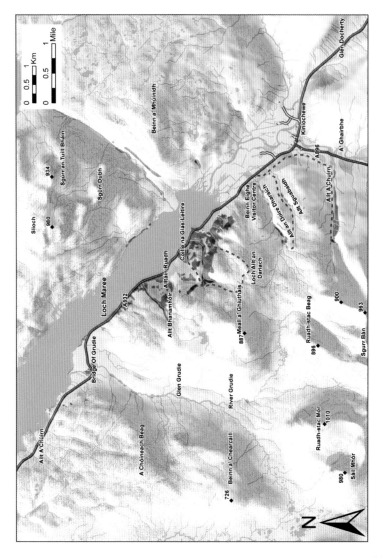

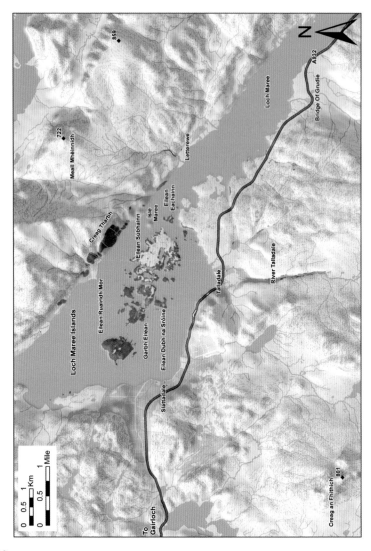

Nature reserve trail at Beinn Eighe

Travel Notes

The nearest railway station is Achnasheen, on the Kyle line, served four times a day by train from Inverness. From the railway station the A832 road leads through Glen Docherty to the village of Kinlochewe. A few kilometres further up the road is the Beinn Eighe Visitor Centre (NH 019 630). The Scottish Natural Heritage nature reserve has a number of well marked routes which allow access to Coille na Glas Leitire. In Glen Torridon there are tracks up to the woods in the gorges at Allt Achairn and Doire Daraich.

The 'mountain trail' starting opposite the carpark at NH 001 650 on the A832 leads up the Alltan Mhic Eogheinn to the Loch Allt an Daraich. The return route beside the Allt na h-Airighe offers spectacular views of Coille na Glas Leitire pinewoods before reaching the road 500 m west of the start point. The small group of pine lining the Allt na Doire Daraich can be reached by the 'pony trail', an upland path that leads to the high tops of Beinn Eighe. The path starts at the Visitor Centre and goes south east then south west up the side of the Allt Sguabaidh to the head of the Doire Daraich trees at NG 997 621. The Allt Achairn pinewood is reached by following the path

Loch Maree and Beinn a' Mhuinidh

south east from the Visitor Centre and then curving south alongside the A896 to join the track west up the Allt a Chuirn. With the massive slopes of Sgurr Ban ahead, follow this track upstream to the waterfall at NH 001 609.

The Loch Maree islands can be seen by heading west from the visitor centre on the A832 to the Slattadale car park (NG 888 721).

The Kinlochewe Hotel is a former 19th century coaching inn, offering bed and breakfast with cheaper beds in an adjoining bothy. There are also a number of guest houses in the area including the Old Mill Highland Lodge, a converted water mill. Along the A832 past the visitor centre is a famous old inn, the Loch Maree Hotel. Queen Victoria stayed here in 1877 when she visited the area, and took a trip by boat to see the islands. There is a small basic campsite at Taagan Farm at the southern end of Loch Maree.

Gaelic Place Names

Loch Maree	loch of Ma-Ruibhe (Saint's name)	Alltan Mhic Eogheinn	little burn of the son of Ewen
A' Ghairbe	the rough one	Coille na Glas Leitire	wood of the grey slope
Beinn Eighe	file mountain	Achnasheen	field of the storms
Sgurr Ban	white/fair peak	Meall a Ghiuthais	hill of the Scots pine
Allt Achairn	the burn of the cairn	Eilean Subhainn	island of the berries
Allt Sguabaidh	the little sweeping burn	Eilean Ruairidh Mòr	Rory's big island
Doire Daraich	hollow of the oaks		
Allt na h- Airighe	burn of the shieling/pasture		

COULIN

Map: OS 1:50,000 Sheet No 25
Wood Grid ref: NG 995 557
Access Point: Loch Clair (NH 000 580) or Achnashellach railway station (NH 003 486)

The mature pine trees in the Coulin wood beside the tranquil, reed-fringed Loch Clair provide an enchanting scene in contrast to the bleak surrounding moorland.

Set within the vast Torridon Mountains, to the south west of Kinlochewe, this is a remote and challenging environment. There are three main blocks of woodland along the course of the River Coulin. A strenuous but rewarding walk can be taken through the Coulin pass, an ancient route that leads to Achnashellach railway station, part of the dramatic Kyle of Lochalsh Line.

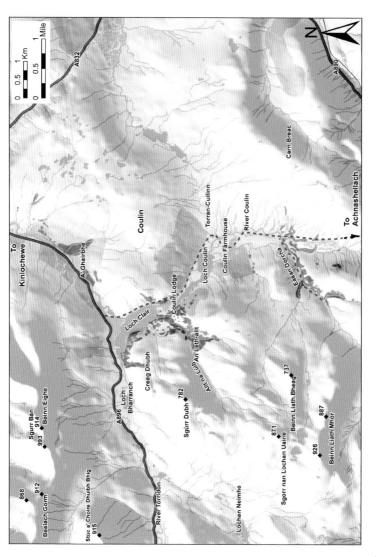

Travel Notes

The forest is situated between Glen Torridon and Glen Carron with the nearest railway station at Achnashellach, another of the Kyle of Lochalsh Line's charming stops. There is an ancient route from the station through the Coulin Pass to Coulin Lodge. This track known as the Old Pony track or Coffin road is part of a long distance cattle droving route from Poolewe and was also used by James Hogg in his travels round Scotland in 1803.

From the station, cross over the railway line, through the gates and head north-east up the steep slopes of the conifer plantation. A good forest track heads east for 2.5 km then joins the old pony track which rises steeply northwards out of the forest. The track continues downhill for 3 km to a stone bridge (NH 023 531). To see the southern pinewood remnant at Easan Dorcha, take the track left going upstream to a small footbridge where the track bends south west and continue for 1 km to the first of the pine trees. Return back down to the stone bridge and follow the River Coulin north, crossing a river plain and along the edge of a plantation to Coulin farmhouse. Turn right over the bridge across the River Coulin past the buildings at Torran

Loch Coulin

A' Ghairbhe pinewood

Cuilinn. A boggy path follows the north shore of Loch Coulin for about 2.5 km then joins a vehicle track that continues north along the east shore of Loch Clair to finally join the A896 at the north end of the pinewood.

Gaelic Place Names

Coulin	collection of enclosures	A' Ghairbhe	the rough one
Torran Cuilinn	hill of the enclosures	Allt na Luib	burn of the bend
Easan Dorcha	the dark waterfalls		

ACHNASHELLACH

Map: OS 1:50,000 Sheet No 25
Wood Grid ref: NH 035 470
Access Point: Craig NH 039 492

The Achnachellach woods consist of mature pine stands amongst an old Forestry Commission plantation. Large scale pinewood restoration is taking place to remove the non-native trees and return the area to pinewood habitat.

Situated a short distance from the tiny Achnashellach railway station the wood lies along three glens which lead into the River Carron. The area is well known for its mountainous scenery with the Munros of Sgurr nan Ceannaichean, Sgurr Chaorachain and Sgurr Choinnich forming a dramatic backdrop to the woods. A long distance hill path leads from Achnashellach to Loch Monar and Strathglass.

Travel Notes

The railway station at Achnashellach has trains four times a day from Inverness. This quaint stop is unstaffed and passengers have to wave a hand to signal the train driver to halt. There is also an irregular bus service.

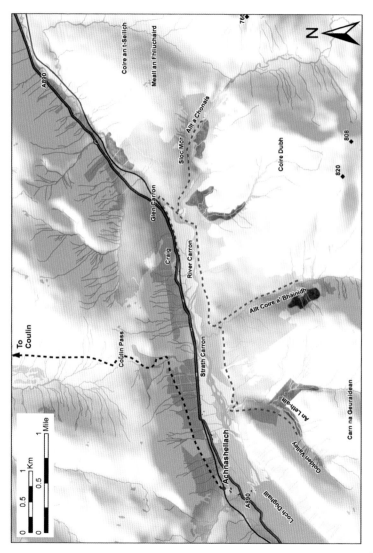

Achnashellach station

From the train station, head down to the main A890 road and go 2.5 km east to Craig. The entrance to Achnashellach pinewood is reached by a track about 150 m past Craig (NH 040 492) which crosses over the railway. The track runs east alongside the railway then turns south crossing a bridge over the River Carron. Ignore the track immediately to the left and follow the major forest road eastwards staying on the north side of Allt a Chonais to reach the rocky gorge at Sloc Mor. Stay on the main forest road and at the next main fork (NH 057 490) go downhill to the right and continue on the forest track for a kilometre. Return along the same route to the River Carron, and just before the main bridge turn left to go across the Allt a Chonais. Continue along the south side of the River Carron for 2 km and just before a small bridge there is a track to the left which leads up Coire a Bhainidh where there are mature pines. Descend and rejoin the main track then continue west for 1.5 km to another small bridge at the start of the Golden Valley (NH 017 481) and the western pinewood remnant.

Gerry's Hostel at Craig is a well known stopover for hill walkers.

Gaelic Place Names

Achnashellach	the field of the willow	Sgurr nan Ceannaichean	peak of the peddlers
Coire a Bhainidh	white/milky corrie	Sgurr Choinnich	moss peak
Allt a Chonais	burn of the gorse	Sgurr Chaorachain	peak of the little field of berries
Sloc Mor	the big gully		
Allt Coire a Bhainidh	burn of the white corrie		

Bridge over the Allt a Chonais

COIR' A' GHAMHNA

Map: OS 1:50,000 Sheet No 19
Wood Grid ref: NH 056 825
Access Point: Corrie Hallie carpark by Dundonnell (NH 114 852)

Tucked below the mighty An Teallach and the buttresses of Corrag Bhuidhe is a small hidden pinewood remnant of less than 100 trees. Pines clinging to the sides of a spectacular deep gorge above Allt na Ghamhna is all that remains of a much larger wood that once stood here up until the 18th century. Now replaced by moorland with scattered dead tree stumps this is an incredible vast isolated area and it is advisable to plan an overnight stay.

Travel Notes

This is a difficult and remote area to reach and is best done with an overnight stay if using public transport.

The nearest train station is in Inverness. There is a bus service on certain days only from Inverness to Dundonnell which takes just under 2 hours. Alternatively there is a daily bus to Ullapool then cycle east along the A835 and take the A832 turning towards Gairloch. There is a parking area alongside the road just before Dundonnell House and a vehicle track on the other side of

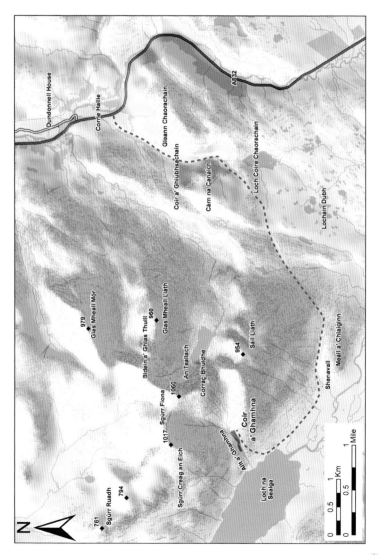

Coir' a' Ghamhna

the road (NH 114 850) beside a set of snow gates, that climbs south west along Gleann Chaorachain. At the top of the hill the track passes Loch Coire Chaorachain on the left and immediately on the right before the track descends, is a path marked by two small cairns. This leads down a steep slope beside the Allt a Chlaiginn to Shenavall bothy. From the western corner of the bothy, by an old rowan tree a narrow path leads alongside a moraine towards Loch na Sealga, past the ruins of an old boat house. Continue along the north shore of the loch rising above a sandy bay till Coir' a' Ghamhna comes into view then follow the fence up the steep slope to see the waterfall.

As well as the Shenavall bothy there is hotel and guest house accommodation around Dundonnell.

Gaelic Place Names

Coir' a' Ghamhna	corrie of the calf	Loch na Sealga	loch of the hunting
An Teallach	the forge	Sgurr Fiona	peak of wine

Descent to Shenavall with Beinn Dearg Mor behind

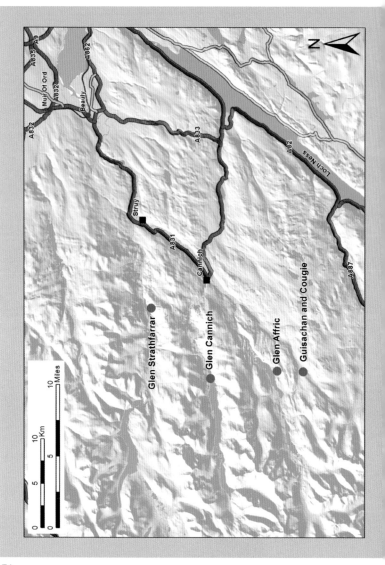

STRATHGLASS GROUP

Strathglass, lies parallel to the Great Glen and from it run three major glens, stretching out to the west. Very large natural forests of pine and birch have long been a dominant feature with several accounts of fine tall trees used for timber. The region has seen much social change as the land was forfeited to the Crown in the 18th century and then the upheaval as people gave way to sheep farming in the 19th century followed by the massive engineering activity of hydro electric developments in the 20th century. Yet the forests have endured and remain hugely important natural features, rich in wildlife.

This is a popular area for visitors to the Highlands with over 100,000 people a year going to Glen Affric, reputed to have some of Scotland's most beautiful scenery, with mixed broadleaved woodland and pine amongst high mountains and lochs. The other glens provide tranquil retreat for walking, cycling and fishing. The RSPB reserve at Corrimony to the south of Cannich is also worth visiting. There is native woodland regeneration alongside open moorland and the chance to see black grouse displaying on a cold spring morning.

The starting point for the pinewoods is the village of Cannich, expanded at the time of the construction of the hydro electric schemes, but now a quiet outpost. The nearest rail link is at Beauly.

- There are buses from Inverness which will carry bikes if booked in advance (see Traveline Scotland).
- Accommodation: Hotels and guest houses in Cannich, Struy and along the glens (see VisitScotland).
- **Pinewoods with major visitor facilities:**
    Glen Affric – Forestry Commission Scotland.

GLEN STRATHFARRAR

Map: OS 1:50,000 Sheet Nos 25 & 26
Wood Grid ref: NH 170 373 to NH 370 390
Access Point: Struy (NH 401 404)

Strathfarrar is the northernmost of the glens stretching west from Strathglass. The large pinewoods extend several miles up this dramatic and picturesque glen once used as a hideout by Jacobite soldiers.

From the starting point at Struy there is a good surfaced road leading through a gate with limited access to vehicles but open for walkers and cyclists. Four large pine and birch woods stretch out along the south of the River Farrar. A good access road leads to the head of the glen and the end of the pinewoods, where there is a large hydro-electric dam.

Travel Notes

Beauly, a few kilometres to the west of Inverness, is the nearest railway station. From here there is a 16 km stretch along the A831 road to reach the entrance of Glen Strathfarrar, on the right before Struy Bridge (NH 401 404). After a kilometre the estate road has a locked gate limiting vehicle access but walkers and cyclists can get through at the side.

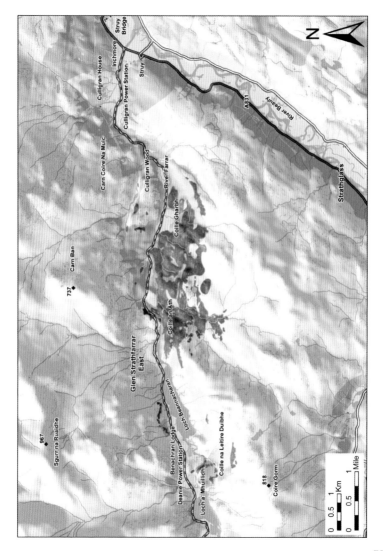

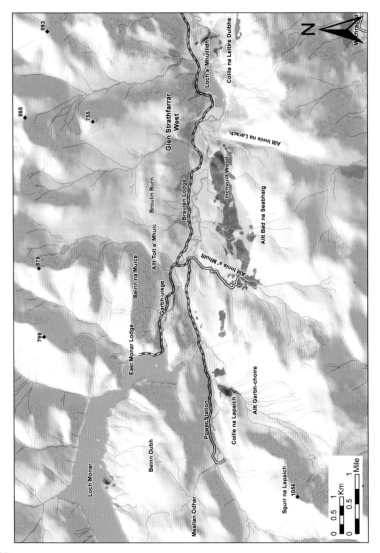

River Farrar

The road is surfaced all the way up to the top of the glen, passing a hydro electric power station and several well maintained Victorian, white painted lodges. The westernmost end of the wood can be seen from the dam (NH 203 393) below Monar Lodge. A track continues along the north shore of Loch Monar and is the route of an old drove road which continues northwest to Glen Carron and the Achnashellach pinewood.

The Struy Inn has a small bar and restaurant with accommodation that welcomes cyclists and walkers. There are also Norwegian chalets and guest house facilities in Struy.

Gaelic Place Names

Glen Strathfarrar	wide winding glen	Loch a'Mhuilidh	loch of the mill
Uisge Misgeach	the drunken water	Coille Garbh	the rough wood
Loch Beannacharan	long tapering loch		

GLEN CANNICH

Map: OS 1:50,000 Sheet No 25
Wood Grid ref: NH 160 300 to 312 325
Access point: Cannich NH 33 31

The ancient pinewoods in Glen Cannich have been decimated by past fires and forest management. They are now under recovery and this beautiful glen is well worth a visit.

A good quality access road leads through the glen past the small groups of remaining pine up to Loch Mullardoch. Large scale forestry work is helping repair the woods by removing non-native conifers to allow the pinewood to regenerate. Despite the upheaval from forestry activity there are tranquil areas and the walk alongside the River Cannich with its small lochs is a rewarding experience.

Travel Notes

There is a bus service from Inverness which stops at Cannich village and also a bike hire facility at Cannich camping and caravan centre. Some of the buses will also carry bikes with prior notice. The nearest train station is at Beauly and from here there is a 20 km journey along the A831 to Cannich.

Entering the village from the north,

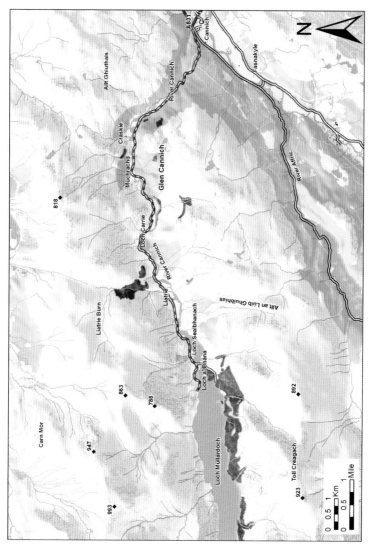

Pine regeneration at Liatrie

cross the bridge (NH 337 318) and immediately turn right then follow this road up Glen Cannich. The tarmac covered road to Loch Mullardoch is relatively easy going. From around Craskie farm (NH 303 337) the pinewoods can be seen on the left stretching east. The road winds round Loch Carrie and at Mullardoch House it splits with one route going to the southern edge of Mullardoch Dam where it stops at a car park. The other goes to the north of the dam and from here a hill path follows the north shore of the loch allowing views of the pinewood on the other side.

Cannich has guest houses, a youth hostel and campsite. The Slaters' Arms is a shooting and fishing themed public house.

Gaelic Place Names

Glen Cannich	glen of the bog cotton	Loch Mullardoch	loch bare rounded promontory
Liatrie	slope		

Pinewood restoration with non native conifers removed

GLEN AFFRIC

Map: OS 1:50,000 Sheet Nos 25 and 26
Wood Grid ref: NH 145 225 to NH 300 285
Access Point: Cannich NH 33 31

Affric is one of the largest and most visited of the pinewoods and a photographer's paradise. Mostly managed by the Forestry Commission the area has a variety of walks to suit all abilities.

The woods stretch over 25 km westwards from Cannich village with the main pinewood on the south shore of Loch Beinn a' Mheadhoin. Well marked walks offer short circular loops in the woods or there are long distance trails into the surrounding hills with spectacular views over the lochs and magnificent pinewoods. From an early stage in its ownership of the woods, the Forestry Commission decided to conserve the pines and to encourage natural regeneration as well as providing excellent visitor access and interpretation facilities.

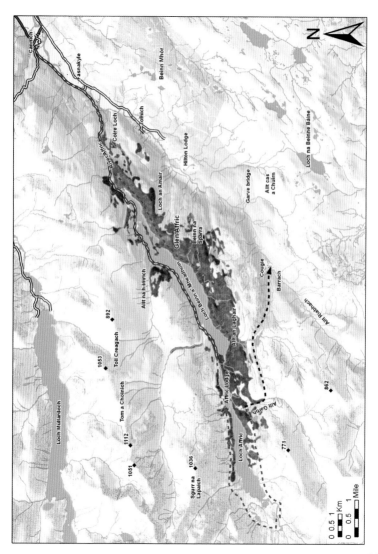

Loch Beinn a' Mheadhoin

Travel Notes

Glen Affric can be reached by public transport with a bus service to Cannich from Inverness. In summer a bus also continues up Glen Affric. Bikes can be carried on some of the buses by phoning first to make arrangements. Alternatively bikes can be hired in Cannich village. From here there is a good road going west, past the Slaters Arms pub. After passing the Fasnakyle hydro-electric power station continue for 4 km then cross the river at (NH 289 283) to take the short Dog Falls loop walk up the steep slope to the lovely Coire Loch and down again to a carpark. Back on the road, go west, alongside Loch Beinn a' Mheadhoin, with its wooded islands. After crossing the small Chisholm Bridge near the end of the loch the road ends at a carpark (NH 201 233).

The walker now has a choice of heading west around the north shore of Loch Affric or crossing the bridge and going round the easier southern route. If going over the bridge, the track comes to a junction. The track on the left goes back east along the shore of Loch Beinn a' Mheadhoin. The right fork leads west up the glen alongside Loch Affric and brings you opposite Affric Lodge and a crossing over the Allt Garbh. There is a faint track to the left, heading south, which leads to Cougie pinewood in the neighbouring glen. Continuing on the main route west brings you to the end of the loch. Follow the river to the buildings at Athnamulloch. Cross the wooden bridge (NH 132 206) and pass the delightfully named Strawberry Cottage, then at the next junction (NH127 208) turn right back along the loch's north shore to the carpark.

Gaelic Place Names

Beinn a' Mheadhoin	middle mountain
Allt Garbh	the rough burn
Loch Beinn a' Mheadhoin	loch of the middle mountain
Coire Loch	loch of the corrie
Glen Affric	Glen of the dappled woodlands

GUISACHAN AND COUGIE

Map: OS 1:50,000 Sheet No 25, 34
Wood Grid ref: NH 230 200 to 298 235
Access point: Tomich NH 307 273

Much of the original woodland in Guisachan forest has been destroyed by past commercial forestry activity but a new conservation programme is encouraging regeneration of the pines.

The pinewood will take time to heal but despite the obvious scars this is a wonderful glen to visit. A small, relatively untouched ancient stand of pines at Barrach Wood near Cougie shows just how rich and diverse this pinewood habitat can be. A good access road leads from the quaint Victorian purpose built village of Tomich and on to well-marked forest trails. There are short walks to dramatic waterfalls in the woods or a longer route up to the head of the glen and over into Glen Affric for the more ambitious traveller.

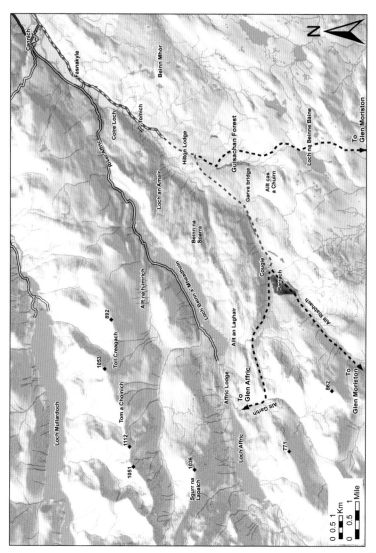

Logs from restoration management at Guisachan

Travel Notes

This remote woodland can be reached by public transport, with a bus service to Tomich or Cannich from Inverness bus station. Bikes can be carried on some of the buses by phoning in first to make arrangements. Alternatively, bikes can be hired in Cannich. From Tomich (NH 307 273) there is a good forest road for about 4 km to Hilton Lodge, a smartly maintained, grand building with a pond full of water lilies in front. A forest road continues alongside the Guisachan pinewoods, west to Garve Bridge (NH 266 222) and then past the farm at Cougie (NH 241 210) where the Barrach Wood lies a few hundred metres to the west.

Tomich has a hotel and bar and there is an independent hostel at Cougie Lodge. There is also Scottish Youth Hostel Association accommodation and guest houses in nearby Cannich.

Gaelic Place Names

Guisachan	place of the pine	Allt Cas a Chuirn	steep cairn burn

Barrach wood

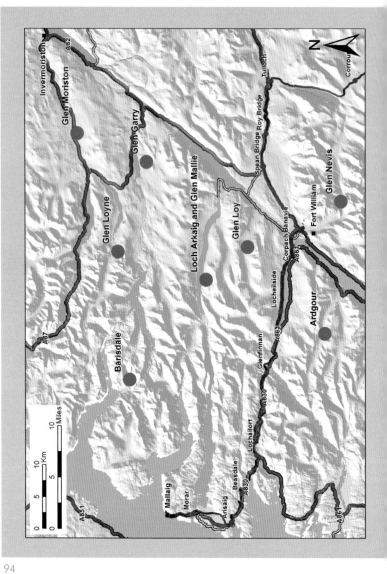

GREAT GLEN GROUP

The Great Glen is a massive geological fault splitting Scotland more or less diagonally in two, with several glens forming branches that head off west. Scotland's highest mountain, Ben Nevis, is at the south-west end near Fort William. This busy town is also one terminus of the Caledonian Canal, which provides a maritime link between east and west Scotland. Mostly, these days, the canal is used by pleasure craft. This wild, mountainous terrain provides ancient access to the Isles and was the escape route used by Bonnie Prince Charlie after his Jacobite army was destroyed at Culloden. Many of Scotland's great fishing rivers and lochs lie in the adjacent glens, often significantly modified by the hydro-electric developments of the mid 20th century. This group of pinewoods also includes Barisdale on the Knoydart peninsula, part of what was known as the Rough Bounds and accessible only by boat or by a 25 km walk along hill tracks. This is a landscape of vast open moors, Munros and forests. It is the perfect place for watching moorland grouse and waders as well as birds of prey which have occupied this land for thousands of years and whose rich abundance has been remarked upon by travellers over the centuries.

- Inverness, Fort William and Spean Bridge provide the most convenient railway stops and the glens are well served by buses (See Traveline Scotland)
- The Great Glen Way runs from Fort William to Inverness providing quiet roads and off-road access for walkers and cyclists.
- There is a wide range of accommodation with plentiful campsites, youth hostels, guest houses and hotels including several opportunities to stay in old castles (see VisitScotland)
- **Pinewoods with major visitor facilities:**

    Glen Moriston, Dundreggan Estate – Trees for Life
    Glen Nevis –The Highland Council (NB the main pinewood is on a rough track in difficult terrain).

GLEN MORISTON

Map: OS 1:50,000 Sheet No 34
Wood Grid ref: NH 310 120
Access Point: Fort Augustus (NH 37 09) or Torgyle Bridge (NH 30 12)

Steeped in the history of Bonnie Prince Charlie, Glen Moriston has a wonderful mixture of mountain, river and woodland features. The small remnants of the ancient pinewood are being augmented by new native woodland planting to create a new wild wood.

The old pinewood at Inverwick lies beside the lovely stone built Torgyle Bridge on the south of the river. To the north is Dundreggan forest where 'Trees for Life' are helping restore old Caledonian forest habitat. There are good forest trails in both the old and the new woods. Despite the upheaval from electricity pylon works it is well worth the steep climb up the 18th century military road from Fort Augustus to get close to some ancient pine and dramatic views of the surrounding mountains.

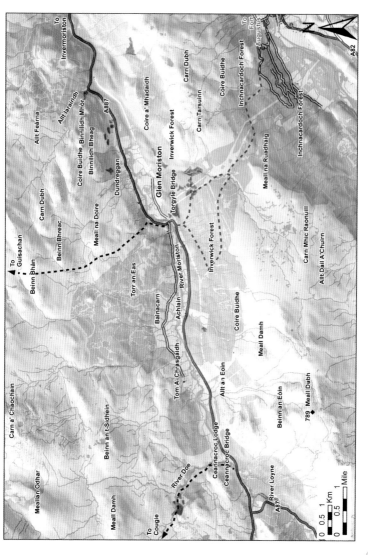

Start of old military road from Fort Augustus

Travel Notes

Glen Moriston is situated midway between the railway stations at Spean Bridge in the south west and Inverness in the north east. From either direction there is a cycle ride of around 50 km to Invermoriston at the mouth of the glen. The Inverness to Skye bus also goes through Glen Moriston.

From Inverness the Great Glen Way provides a walking route to Invermoriston using quiet minor roads, forest tracks and off-road routes which are also suitable for cycling. From Invermoriston follow the A887 to the Torgyle Bridge. Continue over the bridge and along the road, passing the turning to Inverwick Lodge and after a short way there is a large track on the left and small parking space, signposted as the 'Old drove road to Fort Augustus'.

From the rail station at Spean Bridge head for Fort Augustus by taking the A82 north for a short way then turn on to the B8004 across the Caledonian canal to Gairlochy. Join the Great Glen Way turning right onto the B8005 heading alongside Loch Lochy to Clunes and onto the forest track in the South Laggan Forest to Laggan Locks. Cross the Caledonian Canal and travel northwards on the east side of Loch Oich, then alongside the canal into Fort Augustus.

From Fort Augustus, the shortest route to Glen Moriston is 15 km by the 18th century military road that climbs up the steep slope of Inchnacardoch forest to Torgyle Bridge. Starting at the route of the Great Glen Way on the A82 (NH 379 094), take the residential road west towards Jenkins Park. As the road bends sharply left take the turning right beside a telephone box and continue through a residential area till the road becomes a track. After a few hundred metres there is a path to the right (NH 368 095), signposted to Glen Moriston, that zig-zags uphill through old birch trees and bluebells. This path continues through the Inchnacardoch forest heading west then northwest, to become a major access track for the upgrading of the power lines. At the crossroads continue straight on keeping the pleasant rowan fringed Allt na Fearna to the left then cross over by a bridge and under the power lines. Ignore the track to the right and the fork downhill to the left and continue straight on for another 1.5 km. At NH 325 109 there is the option of continuing along the main access road down to Torgyle Bridge or avoid the construction works by taking the small path on the left which is the original military road heading down to cross the Allt Phocachain. Continue west along the military road for another 3 km and before crossing the Allt Dubh take the forest track to the right (NH 292 116) and follow it north east down to Torgyle Bridge.

The alternative and longer route from Fort Augustus follows the forest track of the Great Glen Way, north east towards Invermoriston. The track surface is good but there are several stream crossings with steep descents and ascents.

At the foot of the glen is the Glenmoriston Arms, originally an 18th century Inn, frequented by cattle drovers and more recently was reputedly a hideaway holiday retreat for the actor Charlie Chaplin.

Fort Augustus is well equipped with accommodation and is worth a visit for the views of the Caledonian Canal and its lock gates. The Loch Ness youth hostel beside the A82, is 5 km north east of Invermoriston. The Redburn café at Dundreggan, along the A887, is a great resting stop and the garden is a haven for birds, which the owners attract with feeders.

Inverwick Forest with clearings for powerlines

A Route From Glen Moriston to Guisachan and Cougie

From Torgyle Bridge (NH 309 128) go east along the A887 road for a short distance to the new access track for the electricity power lines. This leads northwards along an old route which now follows the power lines all the way up the steep slope to the ridge at Beinn Bhan. Continue down past Loch na Beinne Baine and into Guisachan Forest. From the gate in the deer fence (NH 290 228), follow the forest track north through the woods alongside Allt nam Fiodhag down to the main road along the glen, at Hilton Cottage. Follow the road north east past the pond at Hilton Lodge for 4 km to Tomich (NH 309 274).

A difficult but more tranquil option is to head west from Torgyle Bridge along the A887 to Ceannacroc Bridge and take the track up the west side of the River Doe to NH 198 134 where there is a ford across the river. Follow the hill path up the side of Meall Damh and down towards the Creag Bhog then head north east along Allt Riabhach to Cougie and then continue east to Guisachan and Tomich.

Gaelic Place Names

Allt na Fearna	burn of the alders	Meall Damh	hill of the stag
Allt Dubh	black burn	Allt nam Fiodhag	burn of the bird cherry
Ceannacroc	head of the hillock	Loch na Beinne Baine	loch of the white mountain
Allt Riabhach	brindled burn	Glen Moriston	glen of the river of great waterfalls
Creag Bhog	the bog rock		

Torgyle Bridge

BARISDALE

Map: OS 1:50,000 Sheet No 33
Wood Grid ref: NG 890 030
Access Point: Barrisdale Bay (NG 86 04)

Barisdale is a small pinewood in one of the most remote areas of Scotland with no road vehicle access making access challenging but there is a long and highly rewarding route in for experienced walkers.

The pinewood lies scattered along the shores of Loch Hourn with the main woods in Glen Barisdale deep in the Knoydart Peninsula. Access is either a tough walk alongside Loch Hourn or a shorter but still demanding walk from Inverie where there is a ferry to Mallaig. An overnight stay is advisable but there is plenty choice with a bothy and camping at Barrisdale Bay. The John Muir Trust has begun restoring native woodland in the moorland to the west of Glen Barisdale.

Travel Notes

This is the least accessible of the pinewoods. The nearest train station is at Mallaig and from here a small ferry crosses to Inverie on the Knoydart Peninsula. From the pier, turn right along the main village

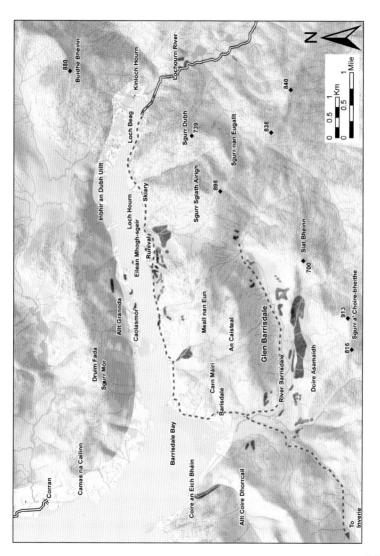

Barisdale Bay

road to Inverie House and then take the track left at NM 773 996 through a woodland. The route continues round to the north side of the Inverie River and after 3 km passes Loch an Dubh-Lochain. The track becomes narrower and there is a steep climb for 4 km to a high point around 400 m at Mam Barrisdale. From here the track descends another 4 km down to Barrisdale Bay at the head of Glen Barrisdale. Continue past the cottages at Ambraigh and cross over the bridge, then take the track eastwards on the north side of the River Barrisdale, with the pinewoods on the opposite slopes.

The alternative route is from Spean Bridge railway station, over 50 km away. From the station follow the A82 north for 1.5 km then turn left along the B8004 and head for the crossing over the Caledonian Canal at Gairlochy (NN 176 841). Turn right, heading north east on the B8005 alongside Loch Lochy till reaching Clunes where the road turns west. At this point, take the track right (NN 200 886) to enter the Forestry Commission South Laggan Forest and follow this route (part of the Great Glen Way) up the western side of Loch Lochy. Depart from the Great Glen Way where it turns off east to Laggan Locks (NN 282 963) and stay on the forest road up the west side of the Caledonian Canal till it joins the A82. Take this road northwards for 100 m then turn left onto a forest track (NN 300 987) to Invergarry. Continue west along the A87 to a small bridge at (NH 242 028) where there is a fork left onto a surfaced minor road. Follow the shoreline of Loch Garry and after 8 km pass the former Tomdoun Hotel. Keep on this road for 25 km, past Loch Quoich and cross the bridge over the flooded Glen Quoich, with a steep descent for the last kilometre down towards Kinloch Hourn. There is a car park beside a small seasonal teashop and farmstead (NG 950 066) where bikes can be left, as the next stage is best done on foot. The road leads down to the jetty at Loch Beag where it becomes a track that runs alongside the south side of Loch Hourn. Although only 12 km to Barrisdale Bay, it is tough going and involves several steep descents and ascents across the rivers that run into the Loch. The route can be very boggy in places but is mainly a stone surface.

There is a youth hostel and a hotel at Inverie. Barrisdale Bay has a basic bothy and campsite as well as a self catering cottage which can be booked in advance. For those travelling past Loch Hourn there is accommodation and a small tea room available for most of the year at Kinlochhourn Farm.

Gaelic Place Names

Ambraigh	the uplands	Loch an Dubh-Lochain	loch of the small black loch

GLEN LOYNE

Map: OS 1:50,000 Sheet No 33
Wood Grid ref: NH 085 050
Access Point: Old Military Road beside Allt a Ghobhainn (NH 112 018)

A small group of scattered trees is all that remains of a once larger forest that occupied this isolated glen. This challenging environment is for experienced hill walkers only.

Far from public transport and set in mountainous terrain this is a difficult pinewood to reach but the vast isolation lends its own dramatic attraction. The start of the journey is on a good access road in Glen Garry. An 18th century military route, which requires care as it often is no more than a sheep track, leads to Shiel Bridge and passes the pine trees below the dramatic rock face of Spidean Mialach.

Travel Notes

Glen Loyne is very difficult to reach. The nearest train station is at Spean Bridge, over 50 km away. See Barisdale for directions to Invergarry. The route to Glen Loyne continues west along the A87 to a bridge at (NH 242 028) where there is a fork left onto a surfaced minor road. Follow the shoreline of Loch Garry and after 8 km pass the former

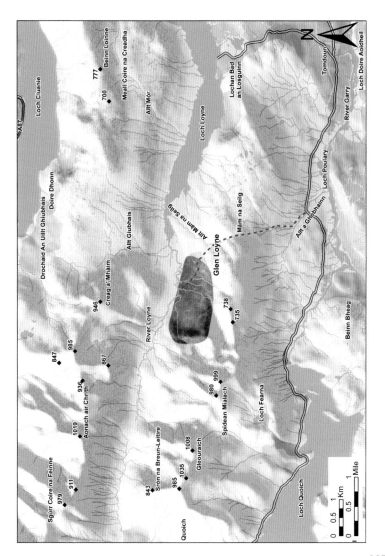

Scattered trees below Spidean Mialach

Tomdoun Hotel. After 4 km the road crosses a small stone bridge at Allt a Ghobhainn (NH 112 018) where bikes can be left, as the route from here is steep and boggy. Just before the bridge is a path heading north west alongside the Allt a Ghobhainn. The start of the route is signposted to Shiel Bridge along the 18th century military road. The steep path heads north towards the western edge of Loch Loyne and rises over 400 m in 2 km. After passing over the saddle at the summit of Mam na Seilg, the path heads north west down towards the River Loyne. In places, the path is hard to find and a compass is advisable. As the path heads downhill, the pinewood can be seen scattered around the dramatic northern slopes of Spidean Mialach.

Spean Bridge has several guest houses and hotel accommodation. In Glen Garry, grand Victorian lodge style rooms can be found at Invergarry Hotel.

Gaelic Place Names

Glen Loyne	beautiful glen	Spidean Mialach	peak of deer or other wild animals
Allt a Ghobhainn	burn of the smith		

GLEN GARRY

Map: OS 1:50,000 Sheet No 34
Wood Grid ref: NH 230 010
Access Point: Mandally (NH 300 007)

The pinewoods fringe the south shore of the large and beautiful Loch Garry in an idyllic setting. There are well marked trails and an old drovers' inn at the mouth of the glen for weary travellers.

The Forestry Commission manage the Glen Garry pinewood much of which had been felled and replanted with alien conifers. The site is now being restored back to pine in a large scale restoration project. Near the White Bridge at the eastern end of Loch Garry there are forest walks which lead to spectacular waterfalls with several more further west at Laddie Burn.

Travel Notes

The nearest railway station is at Spean Bridge, 25 km distant. See Barisdale for directions to Invergarry. The route takes the old road from Mandally (NH 300 007) west to Greenfield (NH 201 005) on the south of Loch Garry. From Easter Mandally the road heads west for 2 km to a carpark. Cross over a bridge and take the track west passing

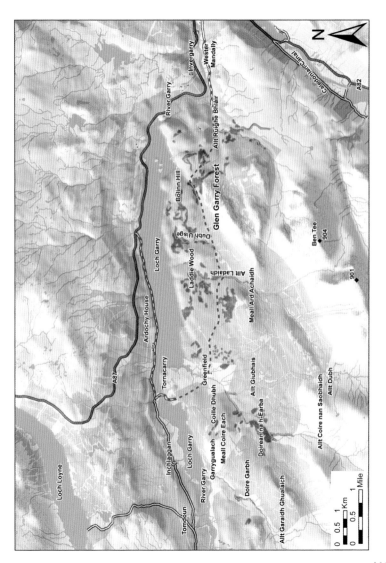

Remnant pines after removal of non-natives in Glen Garry

scattered groups of pine among cleared woodland. Mature pine can be seen where the track crosses the Dubh Uisge (NH 239 006) on the north facing slopes of Meall Ard Achaidh (NN 219 999) and to the east of Greenfield. For an alternative return route, continue west over the Greenfield Burn bridge and take the track northwest down to Torr na Carraidh. Cross the bridge over Loch Garry and at the road turn right back to Invergarry.

Invergarry Hotel is a famous Victorian Coaching Inn with an open fire in the bar and a restaurant. The Glengarry Castle Hotel is nearby.

Gaelic Place Names

Glen Garry	copse glen	Meall Ard Achaidh	hill of the high field
Meall an Tagraidh	hill of dispute	Dubh Uisge	the black water
Torr na Carraidh	hill of the standing stone		

Loch Garry

LOCH ARKAIG AND GLEN MALLIE

Map: OS 1:50,000 Sheet Nos 33, 34, 41
Wood Grid ref: NN 170 875 to NN 010 910
Access Point: Clunes near Gairlochy (NN 20 88)

The vast mountains rising up from the huge Loch Arkaig makes for a dramatic backdrop to the pinewoods set into the north facing slopes.

The Clan Cameron museum at Achnacarry is a good starting point to discover the rich history of the Jacobite uprisings in this area. There is a good access road which leads up the side of Loch Arkaig giving great views of the pinewoods. The adjoining Glen Mallie is more for the hill walker and starts with a rather alarming upgraded forest road which soon leads onto a more typical hill trail. The pinewood bear the signs of a major fire over 70 years ago with bleached dead stems covering the hillside. The Eas Chia-aig waterfall and the witches cauldron below Chia-aig Bridge at the east end of the Loch Arkaig are worth a visit.

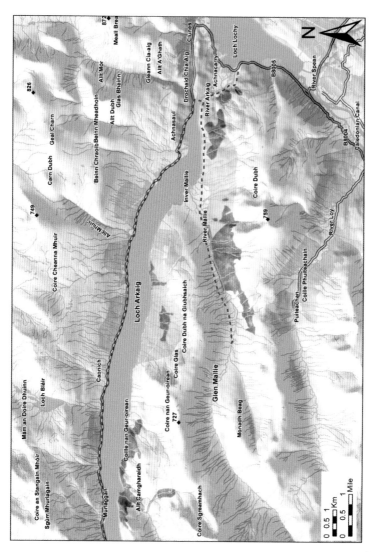

Loch Arkaig

Travel Notes

The nearest train station is Spean Bridge. From here take the A82 north for 1.5 km before turning left along the B8004 heading for the crossing over the Caledonian Canal at Gairlochy (NN 176 841). Turn right on the B8005 and follow this road for about 3 km before turning left onto a side road at (NN 183 871) marked by a sign to the Clan Cameron Museum. After 1.5 km, the road reaches a clearing in the woods where the museum sits. Continue on past Achnacarry House and reach a white painted bridge, which is suitable for walkers and cyclists, to reach the north shore of Loch Arkaig.

The surfaced estate road then follows the glen, west along the north shore of the loch with the Arkaig pinewood on the opposite side.

The track to Glen Mallie leaves Achnacarry House and forks off to the left before passing the white bridge over the River Arkaig. The track has been widened and covered with loose stone to carry forestry vehicles and can be rough to cycle over. The new route now bypasses the bothy at Inver Mallie and crosses the River Mallie at (NN 129 885) then continues west up the glen where the trees can be seen on the south facing slopes. The track ends at a ford in the river near the ruins of Glen Mallie Lodge where the western edge of the pinewood can be viewed.

There are guest houses at Gairlochy and in Spean Bridge.

Gaelic Place Names

Loch Arkaig	loch of the small trout	Drochaidh Chia-aig	bridge of the spray
Glen Mallie	glen of the bare summit	Inver Mallie	mouth of the River Mallie

Glen Mallie with dead tree trunks after 1940's fire damage

GLEN LOY – COILLE PHUITEACHAIN

Map: OS 1:50,000 Sheet No 41
Wood Grid ref: NN 095 840
Access Point: Glen Loy Lodge NN 145 819

Glen Loy, situated north of Fort William and west of the Caledonian Canal offers a great variety of woodland, riverside and moorland features.

The small pinewood at the head of the glen has a wonderful steep mountain backdrop and is varied in age structure unlike some of the more degraded pinewoods. A good road leads up the glen alongside the River Loy. The waymarked forest trails in some of the nearby Forestry Commission oak woods are well worth exploring. The glen is well known for its butterfly populations including chequered skipper, whose colonies can be seen best in June alongside grassy burns.

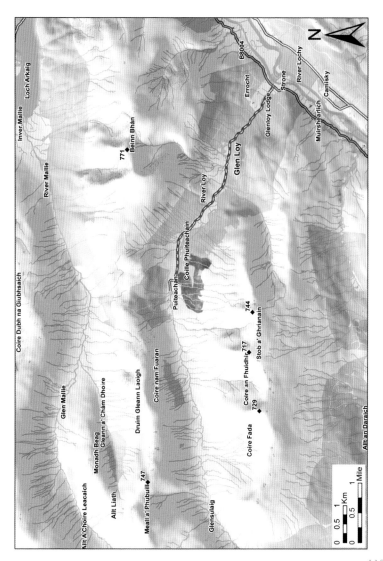

Tunnel under the Caledonian Canal

Travel Notes

The nearest train stations are at Fort William and Spean Bridge.

From Fort William station follow the Great Glen Way past Inverlochy Castle, around the peninsula and on to Neptune's staircase, beside Banavie railway station. The off-road route follows the east bank of the canal and after 8 km reaches the point where the River Loy flows under the canal at (NN 149 817). A track drops to the river goes through a tunnel under the canal and rises on the other side to join the road which continues alongside the river past Glen Loy Lodge. The eastern edge of the pinewood is reached after about 3.5 km. To enter the wood, continue for 3 km till the river and the road bend sharply north. There is a small car park at NN 095 846 at the western boundary of the pinewood. Take the track to Puiteachan farm which crosses the river and immediately turn left then after a few metres a stile goes over the fence into the wood. From here there are no trails so care is needed.

From Spean Bridge Station take the A82 north for 1.5 km then left along the B8004 and head for the crossing over the Caledonian Canal at Gairlochy (NN 176 841). Turn left along the B8004 and after crossing the Bridge at Moy continue 2 km. Turn right after crossing the bridge over the River Loy (NN 147 818) and follow the single-track road alongside the river.

Glen Loy Lodge provides accommodation and guided wildlife tours.

Gaelic Place Names

Glen Loy	glen of the calf	Errocht	place of assembly
Coille Phuiteachain	wood of the swelling knoll	Stob a' Ghrianain	mountain of the sunny hillock

Glen Loy Caledonian Forest Reserve

GLEN NEVIS

Map: OS 1:50,000 Sheet No 41
Wood Grid ref: NN 165 684
Access Point: Carpark at Achriabhach (NN 145 683)

Nestled among the foothills of Scotland's highest mountain Ben Nevis, the pinewood is a small remnant but set in such dramatic scenery it is a worthwhile place to visit.

The trails are well marked but require great care as they wind above the torrential waters of the River Nevis. From the Visitor Centre at Achintee it is a long and steep climb up to the higher reaches of the glen where the pines can be seen on the steep slopes near the dramatic Steall Waterfall.

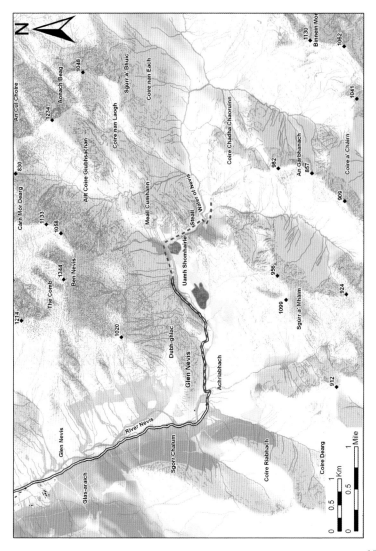

Wire bridge over Water of Nevis at Steall

Travel Notes

From the railway station at Fort William the route follows the A82 to the roundabout at Nevis Bridge. Take the second exit to follow the south side of the River Nevis. Continue along this road for 8 km with some very steep sections before coming to a car park at Achriabhach (NN 145 683). From here a single track road continues another 2.5 km to a car park where bikes should be left. A path winds round the north side of the fast flowing Water of Nevis. Considerable care is needed on this stony wet route. After a short while the woods can be seen alongside the steep gorge and on the opposite slopes of Sgurr a' Mhaim. For the strong nerved there is a simple wire bridge over the river at Steall (NN 177 684) near the spectacular Steall waterfalls, which have the second highest drop in Scotland. Return is along the same route to Fort William.

As a famous venue for visitors the area is well supplied by hotels and guest houses around Fort William and the lower reaches of Glen Nevis. There is a campsite at Achintee beside the visitor centre. The Ben Nevis Inn, a converted stone barn, is a popular traditional old pub with a bunkhouse.

Gaelic Place Names

Glen Nevis	glen of the biting cold water	Steall	the torrent
Achriabhach	brindled field	Sgurr a' Mhaim	peak of the large rounded hill
Achintee	field of the stormy blast	Mamores	the great moors

Path above the Water of Nevis

ARDGOUR

Map: OS 1:50,000 Sheet Nos 40 and 41
Wood Grid ref: NM 960 713
Access Point: Inverscaddle Bay (NN 02 68) or Glenfinnan (NM 89 80)

Set deep in the Ardnamurchan Peninsula, Ardgour provides opportunities for challenging long walks or more gentle strolls that still offer magnificent view of old pines.

The area around Glenfinnan has several Forestry Commission short trails leading into ancient pinewoods. Visitor facilities are provided by the National Trust for Scotland beside the famous viaduct and monument to Bonnie Prince Charlie. The main pinewood in Cona Glen is reached by a long walk along hill trails that offer some spectacular mountain scenery.

Travel Notes

The most convenient train station is Fort William. From the pier (NN 099 738) on Loch Linnhe there is a ferry to Camasnagaul which takes bikes and runs five times a day (no Sunday service). Once off the ferry take the A861 southwest for 9 km alongside the loch to Inverscaddle Bay. Before crossing the bridge, turn right on to the estate road that runs on the north side of the Cona River for about

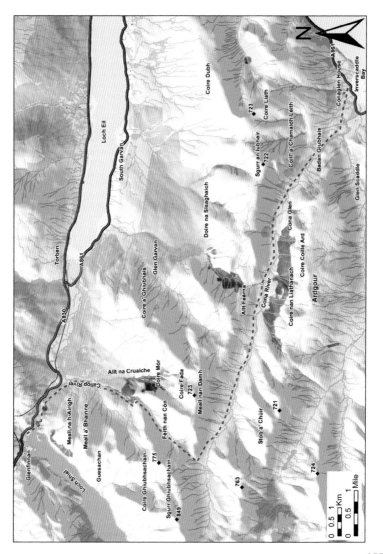

Cona River

6 km till the wood can be seen on the south side.

A more challenging alternative route is from Glenfinnan railway station. From here, head east along the A830 for about 1 km to the road bridge over the River Finnan, with stunning views of the spectacular stone-built Glenfinnan viaduct. At the visitor centre cross the road and head towards the monument then take the first left turn along a track and boardwalk with a wooden footbridge. At the junction with the forest road turn left and go east alongside a forestry plantation, south of the Callop River. After 2 km at a junction, do not cross the bridge over the river, but take the turning right, signposted to Ardgour and continue heading south past the cottages, housing the small Callop power station.

Follow the track as it climbs along the west side of the Allt na Cruaiche towards the old pine trees amongst the conifer plantation at the head of the glen. At the southern end of the woodland a path heads south west alongside the Allt Feith nan Con, below the steep slopes of Sgorr Craobh a Chaorainn for 2.5 km and then turn to the south east towards the Cona River. Continue along the north side of the river for 4 km to the western edge of the pinewood.

Fort William offers many guest houses and hotels. At Glenfinnan there is a static railway sleeper carriage and dining car providing bed and breakfast facilities. This was once a common facility in stations around the Highlands.

Gaelic Place Names

Ardgour	goat-height	Allt na Cruaiche	burn of the heap
Ardnamurchan	the headland of the seals	Sgurr Ghiubsachain	peak of the pine
		Doire Mor	large thicket
Allt Feith nan Con	burn of the bog of the dogs	Meall na h Airigh	hill of the shieling

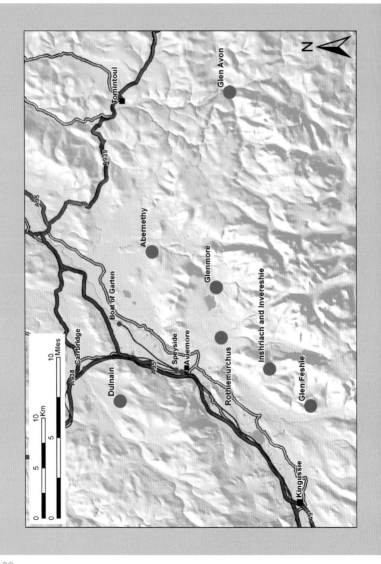

STRATHSPEY GROUP

The River Spey, Scotland's fastest flowing river, lies in the heart of the Cairngorms National Park. This mountainous region is famous for its countryside activities and the town of Aviemore is a popular tourist centre. The area is well connected by the main railway lines north from Edinburgh and Glasgow, with good bus services and large sections of off-road cycle and walking routes. The pinewoods fringe the high mountains in the Cairngorms alongside the Rivers Spey and Feshie and in former times would have covered much of the land from Grantown-on-Spey to Kingussie whose Gaelic origin Ceann a' Ghiùthsaich means head of the pinewood. The area has a spectacular array of wildlife and is the largest tract of semi-natural woodland in Britain with the added spectacle of montane woodland, where dwarf trees grow at altitudes of almost 1000 m. Regeneration and restoration of the ancient pinewoods together with other broadleaved woodland in the National Park could soon link together to form an outstanding example of Caledonian forest which in the not too distant future may stretch all the way round the Cairngorms.

The main pinewoods of Abernethy, Glenmore and Rothiemurchus can have over 200,000 visitors a year, but are large enough to allow a wilderness experience. At the other end of the scale are the remote and smaller woods in Dulnain, Inshriach and the upper reaches of Glen Feshie, right down to the few dozen mature trees in Glen Avon.

- The railway stations at Aviemore, Kingussie and Carrbridge provide convenient stops for the woods as well as local and national buses (see Traveline Scotland).
- The area has many choices for walking and cycling. There is a wide range of accommodation from youth hostels, camping and bothies, to guest houses and hotels (see Visit Cairngorms).
- **Pinewoods with major visitor facilities:**

    Glenmore – Forestry Commission Scotland
    Abernethy Forest – RSPB
    Rothiemurchus – Rothiemurchus estate

GLEN AVON

Map: OS 1:50,000 Sheet No 36
Wood Grid ref: NJ 177 074
Access point: Queens View car park, Tomintoul (NJ 165 176)

A handful of mature pines in a secluded river bank deep in the Grampian Mountains is all that remains of this forest. Conservation work is now helping extend the wood around the Linn of Avon, an attractive water feature 12 km south of Tomintoul.

A hill track heads out from Tomintoul and then follows the River Avon upstream into this remote mountainous area. The starting point at Tomintoul is reputedly one of the highest villages in the Scottish Highlands so this is a route that is best suited for those with moderate hill walking experience. Although the wood is presntly only small this is a wonderful place for experiencing the wild Grampians.

Travel Notes

This is not an easy site to reach by public transport. The nearest train station is at Aviemore. From here take the excellent Speyside Way track (also National Cycle Route 7) north through a housing estate for a short distance then across heather moor alongside the Strathspey

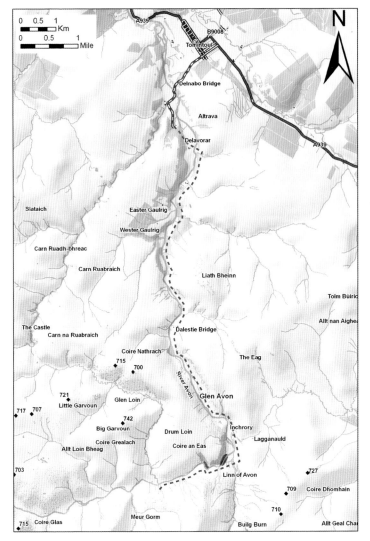

River Avon looking towards Inverloin

Steam Railway towards Boat of Garten. Follow the Speyside Way through the town, turning east across Garten Bridge then turn left along the Speyside Way onto the B970 (NB National Cycle Route 7 departs here and heads south). The route turns right at East Croftmore (NH 960 195) then turns left to go 'off road' at NH 968 191 towards Nethy Bridge. After crossing the bridge over the River Nethy turn right just before the hotel to head eastwards for about 9 km to the junction with the A939 and then turn right towards Bridge of Brown. This road has some steep inclines rewarded by a very fast descent to the bridge where there is a tea room offering a welcome break. Continue for another 9 km with the long climb to Tomintoul, reputedly the highest village in the Highlands. At the far end of the village is a right turn (NJ 171 183) to the Queens View car park and the start of a well marked track to the Linn of Avon waterfalls. The easiest route is to follow the public road

over the Delnabo bridge to the end of the road at Altrava (NJ 162 164) and the start of the estate track which crosses a bridge at Delavorar. The route then goes south alongside the River Avon past the Birchfield farm steading and its impressive black gates, to Inchrory (NJ 179 080). Stay on the track heading south alongside the river for half a kilometre then turn west to cross the wooden bridge (NJ 179 073) over the river to the Linn of Avon waterfall and the start of the pinewood. The track continues through regenerating pine which extends to the bridge near Inverloin.

Tomintoul has a youth hostel, guest houses and a hotel. There are also hotels further down in Nethy Bridge and Boat of Garten as well as numerous budget accommodation options.

Gaelic Place Names

Glen Avon	river glen	Linn of Avon	river water feature

DULNAIN

Map: OS 1:50,000 Sheet No 35
Wood Grid ref: NH 830 180
Access Point: Dalnahaitnach, Carrbridge (NH 89 22)

The pinewoods surround the River Dulnain and stretch over to Kinveachy west of Carrbridge at the foot of the Monadhliath mountains.

This is a popular and accessible area for walkers and cyclists. National Cycle Route 7 leads into the forest. There are easy to follow but strenuous hill walks including the 'Burma Road' which leads to the A9 just south of Aviemore. The woods here are important for a variety of pinewood birds such as Scottish crossbill, capercaillie and crested tit. Across the river from Dalnahaitnach look out for the monument to Little John MacAndrew, a famous 17th century inhabitant of the area.

Travel Notes

From the railway station at Carrbridge take the road that heads south west under the A9 and along the south of the River Dulnain to Dalnahaitnach. This is part of the National Cycle Route 7 and for a

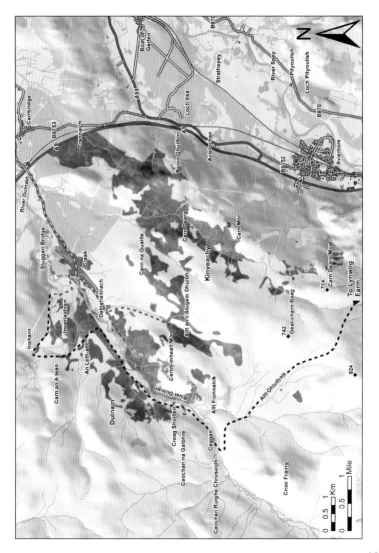

circular trip follow this route west, taking the turn off from the road at NH 878 216 to Sluggan Bridge. This incredible high dome stone bridge replaced the older, two arch structure which was washed away in 1829. Come off Cycle Route 7 at NH 860 218 towards the ruins at Inverlaidnan then head back east, crossing the wooden bridge over the River Dulnain. Join the road heading southwest down to Dalnahaitnach (NH 853 199). A few hundred metres before the old farm building there is a track heading south through a gate in the deer fence. This takes you out on to the open moor of Kinveachy Forest and after 2 km there is a nice group of mature trees beside Allt an t-Slugain Dhuibh. Alternatively, at Dalnahaitnach, a path continues west following a deer fence for a kilometre through some large mature pine trees. Return back along the public road to Carrbridge.

The more challenging alternative is to continue west after crossing Sluggan Bridge along National Cycle Route 7 with the forest plantation on the right to Insharn (NH 842 222). Don't go right here but go through the gate and head south for 5 km to Caggan and then cross the bridge over the River Dulnain along the 'Burma Road', nicknamed for its resemblance to the slow, lengthy original in Southeast Asia. The route

National Cycle Route 7 to Aviemore

Dalnahaitnach

is tough going up the steep hill alongside Allt Ghiuthais and then An Gleannan before going down to Lynwilg farm. Join the A9 and cross over to the B9152 to Aviemore railway station.

Gaelic Place Names

Dulnain	place of the floods	Allt an t-Slugain Dhuibh	burn of the black mud
Allt Ghiuthais	burn of the pine	Dalnahaitnach	haugh of the junipers
An Gleannan	the little glen	Garbh-mheal Mor	the big rough hill

ABERNETHY

Map: OS 1:50,000 Sheet No 36
Wood Grid ref: NH 990 180
Access Point: Loch Garten Osprey Centre (NH 978 183)

The largest of the ancient pinewoods, this famous RSPB nature reserve and the adjacent Dell Wood owned by Scottish Natural Heritage supports one of the greatest examples of near natural forest in Britain. Conservation management is ongoing to help extend and repair parts of the woods that suffered damage in previous centuries.

Abernethy Forest is a wonderful place for quiet enjoyment of the amazing pinewood wildlife. Visitor management is carefully designed to minimise disturbance to sensitive species whilst catering for all abilities of traveller. One of the most popular attractions is the Loch Garten Osprey Centre with a variety of routes from here into mature forest. At Nethy Bridge an information centre provides a good starting point for the route into the old pines in Dell Wood.

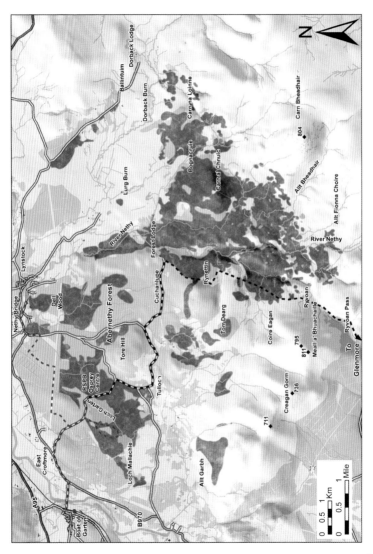

Travel Notes

From Aviemore railway station a connecting steam train goes to Boat of Garten but doesn't carry bikes. The alternative is to join the Speyside Way track (also National Cycle Route 7 in part) at Aviemore and head north through a housing estate for a short way then across heather moor towards Boat of Garten. Follow the Speyside Way through the town, turning east across Garten Bridge then left onto the B970 (NB Cycle Route 7 departs here and heads south). Stay on the main road and turn right at East Croftmore (NH 960 195), signposted to the Loch Garten Osprey Centre. There are many unsigned forest trails and tracks through the forest for those seeking access to explore more widely but a good pinewood experience can be had on the marked paths that go deep into the heart of the mature forest.

From the Osprey Centre go back up the track alongside the north shore of Loch Garten to a carpark and take the circular trail on the left that heads southwest to Loch Mallachie. Once back at the car park turn left and follow the main track for just under a kilometre to join the Speyside Way on the left to Nethy Bridge. Before crossing the bridge over the River Nethy, turn right along Dell road and after 700 m take the track on the right into the Scottish Natural Heritage owned, Dell Wood National Nature Reserve where there is a choice of routes going south into the mature pinewood. In Nethy Bridge, the Explore Abernethy Information Centre is a useful place to learn about the forest and its cultural history.

Basic accommodation is available at Abernethy Bunkhouses and Fraoch Lodge Bunkhouse. The Lazy Duck hostel, one of the smallest independent hostels in Scotland, is a wooden mountain hut with its own wood shed sauna. The Victorians built several grand hotels in the area including the Nethy Bridge Hotel and the Boat Hotel in Boat of Garten. There are also numerous guest houses in this popular tourist area.

River Nethy looking north to Abernethy Forest

Gaelic Place Names

Abernethy	mouth of the River Nethy	Balliemore	the big farm
Dell	the low lying land/haugh	Loch Mallachie	loch of the curse
Ben Macdui	hill of the black pig		

GLENMORE

Map: OS 1:50,000 Sheet No 36
Wood Grid ref: NH 980 090
Access Point: Glenmore Visitor Centre (NH 977 097)

This once large ancient pinewood suffered massive clearance of mature pine in the early 20th century and then was planted largely with non native conifers. This Forestry Commission site is now undergoing a transition as conservation management clears out the alien trees to allow pine regeneration.

Glenmore is well equipped for a variety of outdoor activities including walking, cycling cross country skiing and water sports on Loch Morlich. National Cycle Route 7 and the Old Logging Way provide off-road access from Aviemore. Mature pine trees can be found on the routes from the Forestry Commission Glenmore Visitor Centre around Loch Morlich and also on the track into Ryvoan pass. This old cattle droving route passes the secluded and delightful An Lochan Uaine, 'the green lochan'.

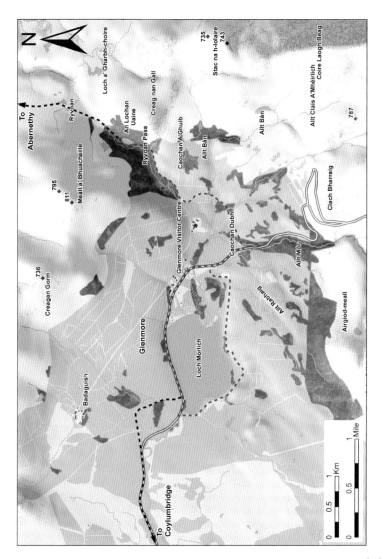

Natural pine regeneration, Glenmore

Travel Notes

Aviemore railway station is close to the forest. From here, head south to join the off-road track part of 'National Cycle Route 7' which heads east, alongside the B970 towards Inverdruie and Coylumbridge. After crossing the River Druie at Coylumbridge the track turns east and becomes the Old Logging Way. This excellent 5.5 km cycle trail goes through spectacular heather moor and mature trees and then along the north shore of Loch Morlich before arriving at Forestry Commission Scotland's Glenmore Visitor Centre. There are several well marked trails from here through the forest. Some of the best areas of native pine can be seen by taking the trail along the south shore of Loch Morlich from the hay field carpark (NH 980 091). Another pleasant option is to head north east to the Ryvoan Pass and An Lochan Uaine. Leave the carpark at the footbridge over the Allt Mor (NH 984 086). Carry on along this forest road over a ford and two footbridges to reach the main track into Ryvoan. Turn right along here for 1.5 km to reach An Lochan Uaine. This track also continues on to Ryvoan Bothy at the southern edge of Abernethy Forest.

Convenient accommodation can be found at the Glenmore Campsite, Loch Morlich youth hostel and Glenmore Lodge.

Gaelic Place Names

An Lochan Uiane	the green lochan	Meall a' Bhuacaille	mound of the herdsman
Ryvoan	slope of the bothy	Loch Morlich	loch of the big hill

Glenmore

ROTHIEMURCHUS

Map: OS 1:50,000 Sheet No 36
Wood Grid ref: NH 920 080
Access Point: Rothiemurchus Centre (NH 902 109)

The second largest of the ancient pinewoods and a favourite tourist destination at the 'gateway to the Highlands' in Strathspey. Some magnificent mature pine trees, remnants of the large fellings during the two World wars, are surrounded by new growth of young trees.

Rothiemurchus is a popular holiday venue and has many well marked routes for all abilities as well as catering for a wide range of visitor interests. National Cycle Route 7 leads from Aviemore towards Inverdruie and the Rothiemurchus Visitor Centre. Several cycling and walking routes offer a variety of short and long distance trails around the forest including the renowned Loch an Eilein and its 14th century castle ruin.

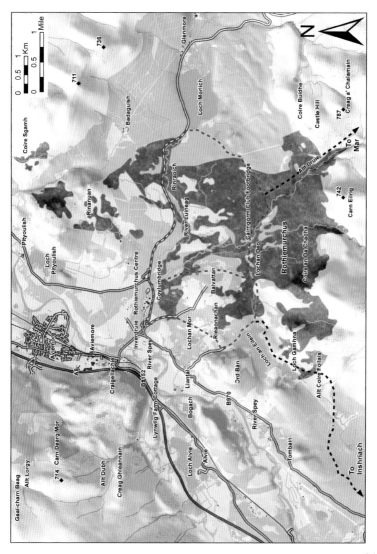

Travel Notes

Aviemore railway station is close to the forest. From here, head south then join the off-road track part of 'National Cycle Route 7' which runs east, alongside the B970 towards Inverdruie and the Rothiemurchus Visitor Centre (NH 902 109).

Leave the visitor centre, where cycling and walking maps can be obtained, and take the Old Logging Way east alongside the B970. After 100 m turn right down the road to Blackpark then at the junction take the road on the right which leads past the old croft house. Cross the bridge over the Milton Burn beside a carpark. Take the track south to a small information centre with toilets and then head east along the top of the loch over a footbridge, to join the main track and then turn right heading south east. The track carries on round the loch and after a kilometre take a left turn at the junction near a large boulder with a wooden bench (NH 905 077). Continue up this track for a kilometre to the junction at Lochan Deo. For a shorter return journey, turn left and head north to the campsite at Coylumbridge and then left on the track alongside the B970 back to Inverdruie. Alternatively, continue east to Loch Morlich, crossing the

Cairngorm Club Footbridge

Cairngorm Club Footbridge then alongside the Allt Druidh for a kilometre where there is a junction nicknamed Piccadilly. Ignore the track to the right which leads up the Lairig Ghru to Deeside and go straight until reaching the estate road leading to Rothiemurchus Lodge. Turn left along this road which leads to the west end of Loch Morlich and across a footbridge over the River Luineag. Turn right at the road and then after a few hundred metres join the Old Logging Way and head back west to Inverdruie.

Rothiemurchus has a campsite as well as several guest houses in the area and in Aviemore. The Old Bridge Inn in Aviemore is worth a visit.

Gaelic Place Names

Rothiemurchus	wide plain of the fir trees	Lochan Deo	small kettle hole
		Lochan Mor	the big loch

INVERESHIE AND INSHRIACH

Map: OS 1:50,000 Sheet No 35
Wood Grid ref: NH 870 030
Access Point: Feshiebridge (NH 85 04)

Much of the pine forest on the lower reaches of the River Feshie is within a National Nature Reserve and some smaller areas of pine stand among plantations in nearby Forestry Commission land.

There is a moderate to tough going hill track leading to the pinewoods on the dramatic slopes of Creag Follais. The wood can be accessed from Feshiebridge or by heading south from Loch Gamhna in the neighbouring Rothiemurchus forest. Hill tracks leading to the adjacent Munros offering spectacular views of the pinewoods. At heights of almost 900 m above sea level the highest altitude pines in Scotland can be found here as stunted and contorted old trees less than a metre tall.

Travel Notes

The main gateway to the forest is from Feshiebridge about 7 km south of Aviemore train station. Take National Cycle Route 7 south from behind the station crossing the River Spey and on to Inverdruie.

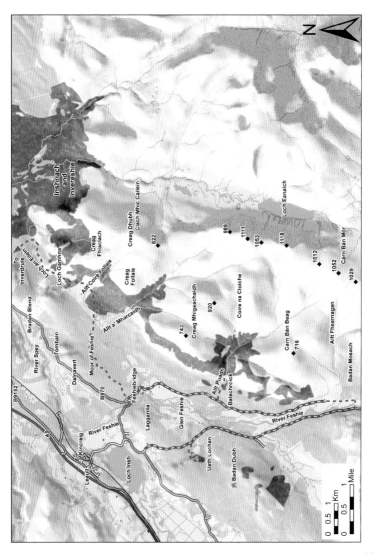

Turn right at the junction to follow the B970 south west. It is well worth popping into the Inshriach alpine nursery along this road. Just north of Feshiebridge, take the minor road east to Lagganlia and Achlean. Go through the plantation forest and then south past the Gliding Club landing strip and continue for a few kilometres. Just before the bridge over Allt Ruadh, there is a pleasant detour up the track which heads east through more plantation and then enters a birchwood before coming to an area of old pine with natural regeneration. Retrace back down to the road, turn left, then cross the stream and continue south passing the Glen Feshie hostel at Balachroick and continue to the cottage at Achlean. The road becomes a rough track which continues south to a flat grassy area past a ruined cottage and heads towards the river where there is a bridge at NN 850 964. Cross the River Feshie here and then turn right to go north down the west side of the glen. Cross the bridge over Allt Fhearnasdail and after 2.5 km there is a track to the left (NH 838 022) to the Uath Lochan carpark and a small group of mature pine. Return to the road then continue north to the junction with the B970 and turn right back to Feshiebridge.

The northern end of the forest can be visited from Feshiebridge by taking the minor road east towards Lagganlia then taking the first forest track to the left and follow this to a junction at NH 857 045. Keep to the left heading north east and after a kilometre turn right at the junction. Continue eastwards for one and a half kilometres then take a right hand turn off the road at NH 878 056 and follow this track east out of the plantation across heather moor over a footbridge and past Drakes Bothy. The track continues north east to Loch Gamhna and meets the Loch an Eilein circular route. Turn right and follow round the Loch to the north end and take the road alongside the Milton Burn up to the B970 then turn right to reach Inverdruie.

There is camping at Glenmore and a Scandinavian style lodge with guest house and self catering rooms at March House beside Feshiebridge. At nearby Kincraig, the rustic Boathouse restaurant offers log fires and food and there is also the Ossian Hotel.

Gaelic Place Names

Invereshie	mouth of the Feshie (river)	Allt Ruadh	the red burn
Inshriach	speckled meadow	Achlean	the broad field
Loch an Eilein	loch of the island	Loch Gamhna	bullock loch
Allt Fhearnasdail	burn of the alder dwelling	Creag Fhiaclach	serrated/toothed crag
		Creag Follais	conspicuous/open crag
Uath Lochan	hawthorn lochan	Badan Dubh	the black copse

Inshriach

GLEN FESHIE

Map: OS 1:50,000 Sheet Nos 35 and 43
Wood Grid ref: NN 845 990
Access Point: Drumguish NN 794 996 or Balnespick NH 837 036

Glen Feshie is renowned as one of the Highlands' most attractive areas. The dynamic River Feshie and the expanse of ancient pine forest has provided inspiration for artists including Edwin Landseer and Hollywood film directors.

Good access roads and paths follow the River Feshie with a number of marked bridge crossings. Hill tracks lead across open moorland into the upper reaches of the Feshie and through the pine trees to the surrounding mountain tops. The area is undergoing a transformation with large areas of pine regeneration as a result of conservation management to control deer grazing.

Travel Notes

The closest railway station is at Kingussie. From here, take the B970 south over the railway crossing then head east at Ruthven Barracks and over the Tromie bridge. Shortly after, there is a minor road to the right to Drumguish. After passing some houses, cross straight over the junction and follow the track east into Inshriach forest. At the

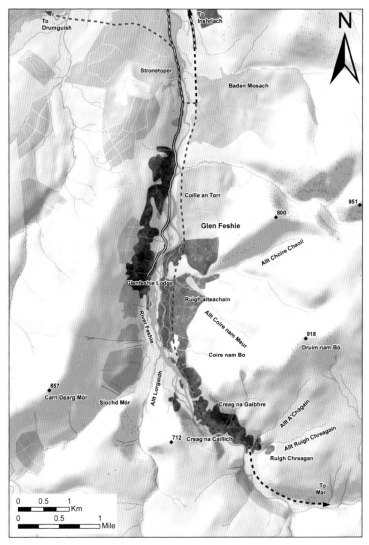

fourway junction go straight on then continue round to the east and exit the forest at Bailieguish ruins where there is a footbridge (NN 823 982) over the Allt Chomhraig. Follow the stream east for 500 m to cross another bridge and continue along the track east, into another block of forestry. Carry on through the trees and straight over at the crossroads as the track heads south east to join the main road along the west side of the glen. Turn right to head south, up the glen, passing the cottage at Stronetoper.

An easier but longer option from Drumguish is to continue along the B970 past Insh Village, to the entrance of an estate road on the right, opposite Loch Insh at Balnespick farm (NH 835 038). Follow the road south, up the glen to Stronetoper. After 500 m turn left at the sign marked 'last crossing over the Feshie'" (NN 849 964), and go over the bridge to the east side of the river. There is a shallow crossing at the Allt Garbhlach where the track goes into the Coille an Torr forestry plantation and then continues through open pasture to the bothy at Ruigh aiteachain. From here on the track can be cycled but the steep slopes below Coire nam Bo are prone to erosion and care is needed in passing. As the River Feshie turns south east, the path reaches lower ground and the end

Glen Feshie

of the main pinewood, although individual trees are scattered right on up the glen. The track continues east along the ancient Feshie Drove Road to Deeside. Return from any point along route this back to the start at Drumguish.

There are guest houses at Insh and several hotels in Kingussie. Feshiebridge has a bunkhouse.

Gaelic Place Names

Glen Feshie	glen of wet boggy river	Allt Chomhraig	burn of the meeting point
Ruigh Aiteachain	the stretch of the junipers	Bailieguish	pine homestead
Drumguish	ridge of the pines	Carnachuin	cairn of the warrior
Coire nam Bo	corrie of the cattle	Creag na Gaibhre	rock of the goats
Coille an Torr	wood of the mound	Creag a Chreagain	crag of the little rock
Stronetoper	nose of the spring	Carn Dearg Beag	the small red cairn
Balnespick	the Bishops farm		

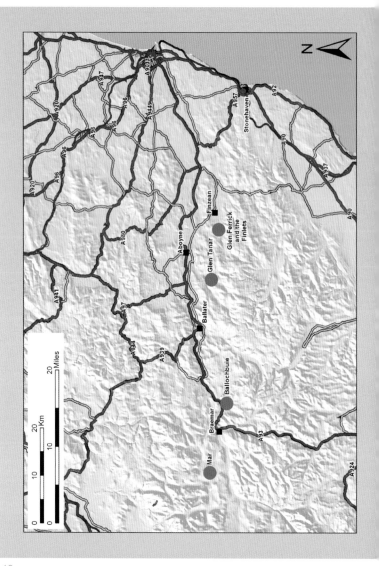

DEESIDE GROUP

With Balmoral Castle at its heart this area has a long history of royal association. Royal Deeside, as it is known, stretches from the Cairngorms and Ben Macdui, the second highest mountain in Scotland, to rolling farmland and the mouth of the River Dee at Aberdeen.

Queen Victoria and the present day Royal family have taken a keen interest in the conservation of the ancient pinewoods here. The Victorian fascination with the stunning romantic scenery is reflected by the large number of grand hotels and quaint decorated cottages in the main towns, from Braemar in the west to Aboyne and Banchory in the east. As well as the four main native pinewoods at Mar, Ballochbuie, Glen Tanar and Birse there are several smaller groups of pine scattered through the area which have naturally seeded or been planted from local stock. Red squirrels are frequently seen in the woods and the elusive Scottish wildcat has a stronghold here along with capercaillie and golden eagle.

The Deeside Railway line was closed in 1966 as part of Dr Beeching's recommendations, although a short 2 km section has been opened up by the Royal Deeside Railway Preservation Society.

- A new cycle and walking route, the Deeside Way has been constructed between Aberdeen and Ballater.
- Accommodation: There are plentiful hotels and guest houses in Deeside.(see Visit Scotland)
- **Pinewoods with major visitor facilities:**

    Glentanar: Braeloine Visitor Centre, Glentanar Estate
    Ballochbuie: Visitor Centre at nearby Crathie, Balmoral Estate

MAR

Map: OS 1:50,000 Sheet No 43
Wood Grid ref: NO 035 932 – NO 085 955
Access Point: Linn of Dee carpark (NO 064 898)

Mar forest on the south eastern slopes of the Cairngorms consists of three pinewoods occupying Glen Derry, Glen Quoich and Glen Luibeg. Owned by the National Trust for Scotland, Mar Lodge estate is being managed primarily for conservation.

This is a dramatic and popular hillwalking area. There are several easy to moderate routes from the carparks at the Linn of Dee and Linn of Quoich to see some of the old pines. More demanding routes for the experienced hill walker lead into the Cairngorms on the old cattle droving roads the Lairig Ghru and the Glen Feshie drove road.

Travel Notes

There is no railway on the east of the Cairngorms so the alternative is to take the bus from Aberdeen to Braemar. A few bikes can be carried, with heavy duty plastic bags provided for free, to protect both bus and bike. From Braemar head west for 10 km, on the road alongside the River Dee, to the Linn of Dee. Just back from the road at Victoria Bridge is the site of

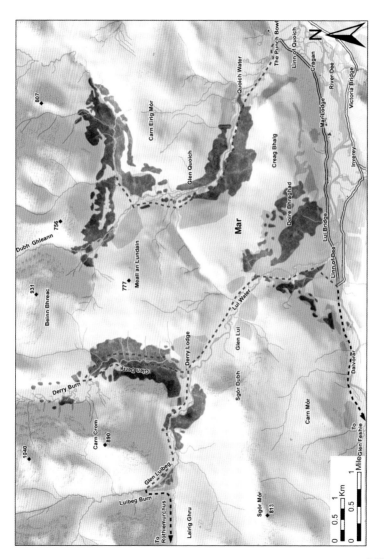

the Gallows tree, an old Scots pine which died in 1920 and was used to hang criminals in the 17th century. For a more demanding journey go by train to Kingussie or Aviemore in Strathspey and then take either the ancient route of the Lairig Ghru or the Glen Feshie Drove road, both of which require stamina and mountain competence.

Aviemore – Lairig Ghru (See Rothiemurchus for Aviemore station to Inverdruie then continue to Coylumbridge campsite at NH 914 106). The track heading south from the campsite is marked by a signpost and there is another one 800 m on at a fork where the route goes left and continues along to the Cairngorm Club Footbridge. Cross over and follow the track to a junction beside a cairn at NH 938 075. Turn right to head south east along the Allt Druidh climbing up to the boulder strewn summit near the Pools of Dee and start the descent past the massive cliffs of An Garbh Coire on the right and the slopes of Ben Macdui on the left. Follow the track down the east bank of the River Dee and at Corrour bothy footbridge, below the Devils Point, take the left fork which goes uphill and round to the east to meet the Luibeg Burn. Cross over the footbridge and follow the Luibeg downstream to Derry Lodge.

Kingussie – Glen Feshie Drove road (See Glen Feshie for Kingussie station to Ruigh Aiteachain Bothy at NN 847 927).

From the Bothy at Ruigh Aiteachain follow the track upstream, on the east of the River Feshie. Be aware of the erosion below Coire nam Bo which may involve crossing loose scree. The track reaches level ground among an open stand of mature pine trees backed by the steep cliffs of Creag na Gaibhre. Continue east along the track to Ruighe nan Leum, where there is a rough boulder crossing over a stream. Another few kilometres on and the track veers northwards up the River Eidart to a spectacular waterfall in a steep gorge crossed by a metal bridge. From here the track crosses open moor and peatbog before reaching drier ground beside the Geldie Burn. The track joins the road from the ruined Geldie Lodge, and continues another 5 km to meet the path from Bynack Lodge ruin. The path heads north east to the

White Bridge over the River Dee. From here a sound but bumpy track follows the river down towards the Linn of Dee car park.

From the car park at Linn of Dee a path heads north and crosses a footbridge to join the main track on the west side of the Lui Water. This then continues north-west towards Derry Lodge passing the remains of the ancient townships where many people lived and farmed up until the mid 1700s. At that time, the estate was forfeited from the Clan chiefs who had supported the Jacobite uprising and the new landowners evicted many of the farmers and their families to make way for timber removal. Later, in the 19th century the remaining population dwindled away as the area became a deer sporting estate. Just past Derry Lodge the track goes west into Glen Luibeg and east into Glen Derry. The entrance to Glen Quoich is 6 km further east of the Linn of Dee with a car park beside the Linn of Quoich. The track runs up the west side of the Quoich water to the pinewoods.

The National Trust for Scotland offer rooms at the magnificent Mar Lodge and also four star apartments, some with Queen Anne four poster beds. Wild camping is permitted but no fires can be lit. Nearby, Braemar has guest houses and a Youth Hostel as well as hotels. The Braemar Lodge Hotel has self catering wooden lodges in its grounds and a small but friendly bar. There are also a number of old Victorian public houses.

Gaelic Place Names

Linn of Quoich	cup shaped water feature	Lairig Ghru	pass of the oozing
Ruighe nan Leum	hill slope of the leap	Glen Luibeg	glen of the little calf

BALLOCHBUIE

Map: OS 1:50,000 Sheet No 43 and 44
Wood Grid ref: NO 200 895
Access point: Old Bridge of Dee, Invercauld (NO 186 910)

Part of the royal estate of Balmoral in Deeside, this large ancient pinewood was spared the 19th century fellings by the intervention of Queen Victoria. Some of the trees here are at least 400 years old and there are lovely areas of peatbog with stunted pine less than 2 m tall and over 100 years old.

The route through Ballochbuie is popular with hillwalkers on their way to the towering cliffs of Lochnagar. From the 18th century Invercauld Bridge 'the old Brig o Dee' there are several walking and cycling routes through the mature pine trees leading onto open moorland with panoramic views of the forest and down to the River Dee.

Travel Notes

There is a bus from Aberdeen to Braemar which takes a few bikes and provides plastic bags to protect them. The entrance to the pinewood is 5 km east of Braemar along the A93. There is a car park at Kelloch on the north side of the road (NO 188 913). From the A93, turn

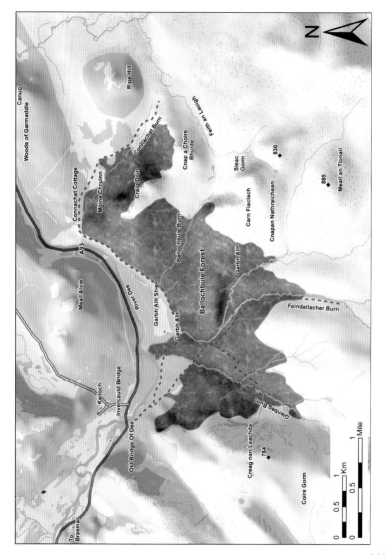

off the road to cross the old Bridge of Dee and then left through a gate in the tall deer fence. Head south east along the forest road for 500 m and take the right hand fork. At the cross roads go straight over to the Glenbeg Burn and on to the Garbh Allt. Take the track right that heads south up the hill to the Garbh Allt waterfalls. There is a pleasant detour from here by continuing up the hill and then taking a left hand turn at NO 197 895 to follow the Feindallacher Burn south to the edge of the pinewood. Return back to the junction and turn left heading southwest to the Glenbeg Burn. Cross over the footbridge and take the right hand fork to head north down the side of the burn to a crossroads. Turn left and go north west to return to the forest gate beside the Bridge of Dee.

A longer route takes in the dramatic east end of the forest. From the Bridge of Dee, enter through the forest gate and south east along the track for 500 m then take the left fork down towards the river and continue to the footbridge over the Glenbeg Burn. Turn right after the bridge then take the next left to cross over the Garbh Allt and follow the track north east to Connachat Cottage (NO 217 918). Follow the track across the bridge and after 600 m take the track to the right at NO 221 924. Continue south east alongside the

Pine below Creag nan Leachda

Southern edge of Ballochbuie

Connachat Burn out to the forest edge. Return is by the same route. There are self-catering cottages available on the royal estate as well as guest houses, hotels and a Youth Hostel in nearby Braemar.

Gaelic Place Names

Ballochbuie	yellow road	Feindallacher	hill of the field of summer pasture
Garbh Allt	rough burn		

GLEN TANAR

Map: OS 1:50,000 Sheet No 44
Wood Grid ref: NO 470 920
Access Point: Bridge of Tanar (NO 480 965)

The boundary of this large pinewood covering over 1000ha remains much as it did on the 17th century Blaeu's Atlas of Scotland. The structure of the wood has however been greatly altered with centuries of felling, fire and grazing. Now much of this protected site is undergoing conservation management including a strict reserve zone in a small area owned by Scottish Natural Heritage.

The forest has a well-developed network of paths and cycle routes from the Visitor Centre at the Bridge of Tanar. Some of these lead up into neighbouring Mount Keen offering spectacular long treks. Each of three main glens in the forest provide a variety of routes alongside dramatic wooded rivers.

Travel Notes

The nearest train stations are over 40 km away to the east, at Aberdeen and Stonehaven. From there it is a cycle south of the River Dee along the B9077 to Crathes then the B974 to Banchory and B976 to Aboyne.

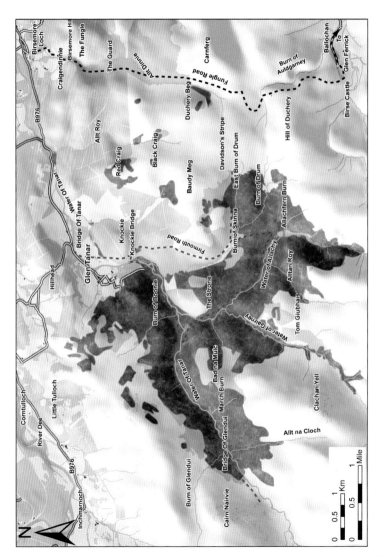

Water of Allachy

Alternatively there is a bus that carries a few bikes from Aberdeen to Aboyne. To reach the pinewoods at Glen Tanar take the Deeside Way, an offroad cycle route west towards Ballater. At Dinnet come off the cycle track and take the B9158 road south, then turn left along the B976 and at the junction with the drinking fountain memorial (NO 471 981) turn right down a minor road to Millfield, which is part of the old drove road known as the 'Firmounth Road'. There is a car park and Visitor Centre near the Bridge of Tanar, at the entrance to the forest.

The forest has a well developed network of footpaths and cycle routes. The best starting point is to go from the Bridge of Tanar south, passing a small chapel built in 1872 and named after St Lesmo, a hermit who lived in Glen Tanar over 1000 years ago.

Continue upstream to the Knockie Bridge viewpoint (NO 480 952). To see the west of the pinewoods, cross the Knockie Bridge then turn left and follow the track alongside the Water of Tanar. Near the edge of the pinewood, the track passes a small wooden shelter 'The Half Way Hut' at the Bridge of Glendui. Return along the track to a bridge over the Water of Tanar (NO 460 941) and cross. Keep left at the fork to cross the bridge over the Water of

Allachy. Turn right to head south east alongside the Allachy and continue along this track as it heads east. After 2 km do not take the fork right over the footbridge but stay left and follow the track as it turns north to join the Firmounth Road. Either turn right here to see the edge of the wood after 1 km or turn left and return through the forest to the Knockie Bridge.

The Glen Tanar estate has a number of self-catering cottages. At Aboyne there are guest houses, hotels and a campsite. The Boat Inn at Aboyne provides excellent meals.

Gaelic Place Names

Glen Tanar	glen of the thundering river	Gairney	rough water
Allachy	little burn	Knockie	small hill

GLEN FERRICK AND THE FINLETS

Map: OS 1:50,000 Sheet No 44
Wood Grid ref: NO 570 915
Access Point: Finzean (NO 60 92)

The ancient pinewoods at Glen Ferrick and the Finlets along the Water of Feugh lie in the Forest of Birse in Deeside. Having once been a royal forest before being conveyed to the Bishops of Aberdeen in the 12th century the area has seen centuries of exploitation. In the 1820s two water powered sawmills were built on the Feugh and there followed considerable felling of the pine trees in Glen Ferrick. Nowadays the woods are being managed for conservation and natural regeneration is abundant.

A long distance walking route to the pinewoods from Aboyne follows the Fungle Road, an ancient cattle droving route reputedly also used by whisky smugglers. A pleasant respite from a steep section of the walk is given at a bench marked 'rest and be thankful' before leading across the moors and into a dramatic steep sided glen. There is also a quiet public road, which passes both woods leading offering a less strenuous route, from Finzean alongside the Water of Feugh to Birse Castle.

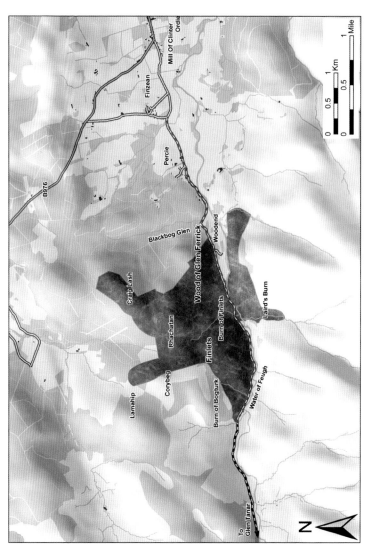

Travel Notes

Aberdeen is the nearest train station over 35 km to the east. The route can also be cycled largely off road along the Deeside Way. From Aberdeen, enter Duthie Park (NJ 938 046) and go west following the line of the old railway to Banchory. From here, take the B974 road south over the Bridge of Dee to Strachan and then the B976 west to Finzean. After 4.5 km, look for the minor road to the left, signposted to the Forest of Birse, beside a large monkey puzzle tree (NO 616 924). As the road turns north through Finzean take the road to the left, signposted Forest of Birse. Glen Ferrick pinewood on the Finzean estate is about 4 km west of the village. Further along the road is the Finlets pinewood which ends by the bridge over Bogturk Burn. Self sown pinewoods also occur on the north side of the remaining 2.5 km of the road leading to the Forest of Birse Kirk at Ballochan.

Alternatively, there is a bus that carries a few bikes from Aberdeen to Aboyne. From here take the B968 south to cross the bridge over the River Dee and join the B976 at Birsemore. Turn right along the road and after 100 m take the turning on the left to the start of

The Finlets

the ancient Fungle Road. Continue on to the western end of Birsemore Loch and follow the road south. Where the road turns sharply right to Craigendinnie, take the track south, signed to Tarfside, and continue until reaching a cottage called The Guard. A footpath goes south from the cottage through the woodland and crosses the Allt Dinnie at (NO 518 949). Where the trees thin out, the path joins an estate track which is the continuation of the Fungle Road. Carry on along this track past a keepers hut and finish east of Birse Castle where there is a car park beside the Kirk. Take the road east alongside the Water of Feugh to see the pinewoods on the left. A farm shop and tea room provides a pleasant rest stop in Finzean.

Gaelic Place Names

Glen Ferrick	glen of squirrels	Birse	thicket
Finlets	the white slope		

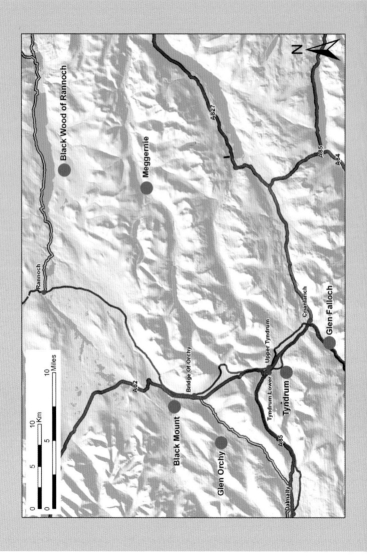

SOUTHERN GROUP

This group stretches from the great expanse of peatbogs at Rannoch Moor to Breadalbane and the southern pinewood limit at Glen Falloch, with part of the area lying in the Loch Lomond and Trossachs National Park. Folklore often describes these southern woods as dark places, lawless and dangerous, with names like the Black Wood and Black Mount. Today they are more likely to be marvelled at as natural spectacles where a real sensation of wilderness can be felt and it takes no stretch of the imagination to step back over the centuries. Several famous writers passed through these woods on their Highland journeys including Charles Dickens, William and Dorothy Wordsworth and even the naturalist, Charles Darwin. They stopped at old Inns such as the Kingshouse and Inveroran, purpose built in the 18th century after the Battle of Culloden along with the major road improvements.

- The West Highland railway line that runs through Crianlarich and splits at Tyndrum, to reach both Fort William and Oban, provides some of the most dramatic train journeys in the world with several pinewoods on route. Buses go to Kinloch Rannoch and Glen Lyon to reach the Black Wood and Meggernie (see Traveline Scotland)
- The West Highland Way provides an offroad walking route passing Glen Falloch, Tyndrum and Black Mount.
- The Forestry Commission Scotland, North Argyll Cycle Routes passes the Glen Orchy woods.
- Accommodation: there are hotels and guest houses in Crianlarich, Tyndrum and Bridge of Orchy and around Kinloch Rannoch and in Glen Lyon, (see Visit Scotland).
- **Pinewoods with major visitor facilities:**

    Black Wood of Rannoch – Forestry Commission Scotland.

BLACK MOUNT

Map: OS 1:50,000 Sheet No 50
Wood Grid ref: NN 285 418
Access Point: Bridge of Orchy NN 297 396

Three woods, Gleann Fuar, Doire Darach and Crannach make up the Blackmount woods near Loch Tulla on the edge of the great Rannoch Moor peatland.

Part of the West Highland Way from Bridge of Orchy leads to the vicinity of the woods with hill tracks leading into the three woods. The surrounding scenery with the loch and high dark mountains is well worth the trek along 18th century military roads. Following in the footsteps of Dickens and the Wordsworths leads to the old inn at Inveroran which instantly transports the walker back 200 years.

Travel Notes

Bridge of Orchy is the nearest railway station. From here follow the West Highland Way west across the old 1750s bridge over the River Orchy and continue through the trees, up and over Mam Carraigh to Inveroran Hotel. Follow the minor road east for a kilometre, along the south shore of Loch Tulla to the Doire Darach woods.

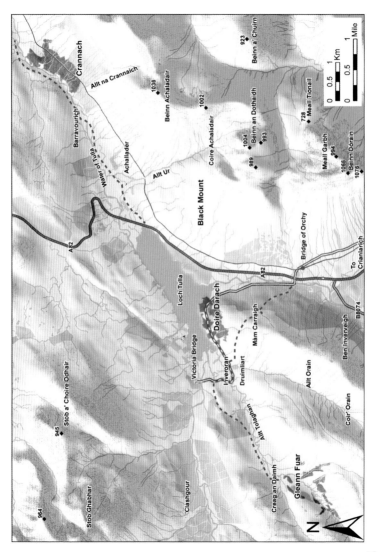

Blackmount – Gleann Fuar

Gleann Fuar by contrast is a difficult wood to reach and involves crossing open boggy moor. From Inveroran Hotel take the West Highland Way crossing a small bridge over the Allt Tolaghan and then shortly after, before reaching Victoria Bridge, there is a car park (NN 270 418) and a small track heading west to a forestry plantation. Just before the plantation fence is the ruin of an old croft and memorial to the poet Duncan Ban MacIntyre who was born here in 1724. Follow the track through the plantation, heading southwest to the fence on the far side beside the Allt Tolaghan. From here, it is possible to pick a way alongside the north bank of the stream, occasionally picking up parts of an old track. Follow the stream for approximately 2 km to the mature pines at Gleann Fuar.

Crannach is reached from Bridge of Orchy by following the A82 north for 5 km and then turning into the Achallader farm track to the carpark (NN 321 442) beside the ruins of Achallader Castle. From the north east corner of the carpark a track heads out east with a ford over Allt Ur, which can be tricky if the water is high. The track continues eastwards and after 1.5 km it crosses a bridge over the Water of Tulla and on towards the old farm building at Barravourich. Before reaching the ruins take the track to the east alongside the water for another kilometre to view the wood stretching out along the lower slopes of Beinn a' Chreachain.

The 18th century Inveroran Inn (now called Hotel) is the nearest accommodation. See also Bridge of Orchy.

Gaelic Place Names

Black Mount	black peatland	Gleann Fuar	the cold glen
Beinn a' Crheachain	hill of the clam shell	Doire Darach	the oak-grove
Barravourich	Murdoch's summit	Mam Carraigh	rounded promontory
Crannach	the place of trees		

GLEN ORCHY

Map: OS 1:50,000 Sheet No 50
Wood Grid ref: NN 250 360 and NN 226 328
Access Point: Eas Urchaidh, car park (NN 242 320).

Two small ancient pinewoods sit like oases in a sea of commercial forest plantations alongside the River Orchy in Argyll and Bute. Owned by the Forestry Commission, the woods are now part of a Caledonian Forest reserve with plans to restore native woodland in the area.

From the carpark near the dramatic Orchy waterfalls well marked forest tracks lead through the plantations to the ancient pinewoods. The route can be challenging with several steep ascents and descents.

Travel Notes

The railway stations at Dalmally and Bridge of Orchy are at different ends of the glen and joined by the B8074 which runs along the east side of the River Orchy. From Dalmally follow the A85 east for a few kilometres to Inverlochy then turn left onto the B8074 and continue for 7 km to the Forestry Commission car-park on the left at the Eas Urchaidh, the dramatic Orchy waterfalls (NN 242 320). Cross over the bridge and

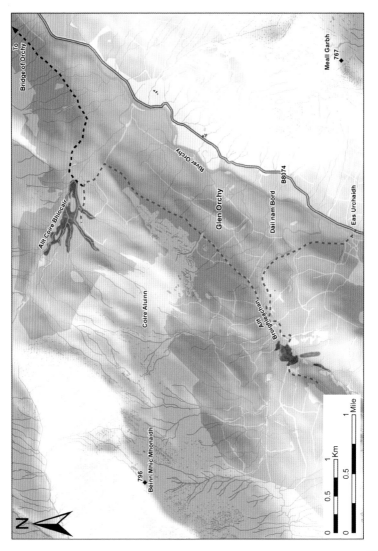

follow the forest track uphill for a kilometre to Allt Broighleachan. The track continues on the south side of the river through dense Sitka plantation for another kilometre and branches right over a footbridge crossing. At this point don't cross the bridge but follow the track south of the river before shortly coming across the magnificent rounded topped crowns of the mature pines in the fenced exclosure. To reach Allt Coire Biochair, go back down to the footbridge and cross over, then follow the track north east through the plantation for about 4 km. There is a ford over the river which needs to be crossed with care if the water is high. From the pinewood in the fenced exclosure here the track continues on another 8 km to Bridge of Orchy. The going can be tough with several steep descents and ascents over stream crossings.

Deer exclosure at Allt Broighleachan

Glen Orchy

The Bridge of Orchy Hotel offers mid to high end facilities. The West Highland Sleeper is a hostel in the original old Bridge of Orchy railway station and provides showers and toilets for campers. The Inveroran Hotel is a renowned old inn long used by famous travellers to this area.

Gaelic Place Names

Glen Orchy	glen of the narrow pass	Allt Broighleachan	tumultuous burn
Allt Coire Biochair	burn of the vicar's wood	Eas Urchaidh	waterfall of the narrow pass

TYNDRUM

Map: OS 1:50,000 Sheet No 50
Wood Grid ref: NN 330 280
Access Point: Dalrigh car park (NN 343 291)

Coille Coire Chuilc Pinewood sits nestled in the foothills of Ben Lui near the village of Tyndrum. It is hard to avoid a feeling of being in a primeval landscape although the truth is that human influence has greatly shaped the area. Much of the forest of pine and oak that once covered the surrounding moorland, was felled to fuel iron foundries and to provide props for lead and gold mines. The most notable feature today is the lack of young trees due to livestock grazing. A conservation area has been established with fencing deep in the wood to help protect the new growth.

Set in the Loch Lomond and Trossachs National Park the pinewood is reached by a short detour off the West Highland Way walking route. There are pleasant walks from either Crianlarich or Upper and Lower Tyndrum railway stations. This is a popular area for hill walkers heading for the Munros of Ben Oss and Ben Lui. Look out for gold panners on the River Fillan and it is well worth a visit to the nearby St Fillan's Chapel.

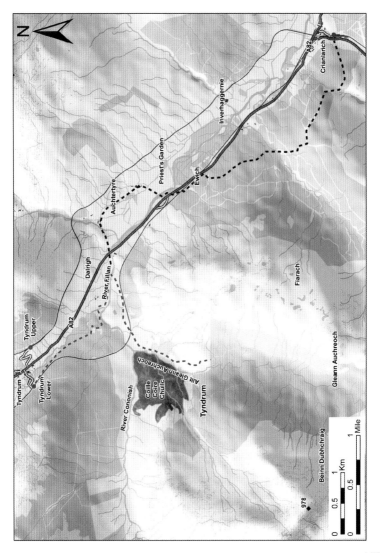

Travel Notes

Tyndrum railway station lies close to the pinewood. There are two stations at Tyndrum, the upper station being the one where trains from Crianlarich go north to Fort William and the lower one for the trains west to Oban. From Tyndrum either cycle east for 1.5 km along the A85 to the car park at Dalrigh (NN 343 291) or walk along the West Highland Way path. At Dalrigh a surfaced road leads to a bridge over the River Fillan. Pass through a wooden gate, cross the bridge and take the track west alongside the railway line. After a kilometre the track crosses the railway by a bridge and heads into the Coille Coire Chuilc pinewood. The track then goes up the eastern side of the Allt Gleann Auchreoch to the southern end of the wood.

Wooden wigwams and canvas yurts are available at Strathfillan Wigwams

Pine tree growing over rock

and the Pinetrees campsite lies a few miles further west. At Tyndrum is a famous old coaching inn, the Tyndrum Inn.

Gaelic Place Names

Tyndrum	house of the ridge	Dalrigh	king haugh
Coille Coire Chuilc	wood of the corrie of reeds	Beinn Dubh	black mountain
		Ben Lui	hill of the calf
Allt Gleann Auchreoch	burn of the speckled farm glen	Ben Oss	hill of the elk

GLEN FALLOCH

Map: OS 1:50,000 Sheet No 50
Wood Grid ref: NN 367 233
Access Point: Crianlarich NN 38 25

Scattered mature pines stand on Dun Falloch and down Glen Falloch, once known as 'the hidden glen'. This is the most southerly pinewood in Scotland. A large forest persisted here up till the 19th century when thousands of trees were felled leaving the 100 or so that we see today. From the 1980s fencing was erected to protect young trees from grazing animals, along with planting of saplings using seed and grafts from the mature pines.

The wood lies on the West Highland Way on the route of the 18th century military road. The Crianlarich hills with seven Munros create an impressive backdrop to the forest. The pinewood is easily reached on the track leading from Crianlarich railway station with its Victorian tearoom and nearby old coaching inn.

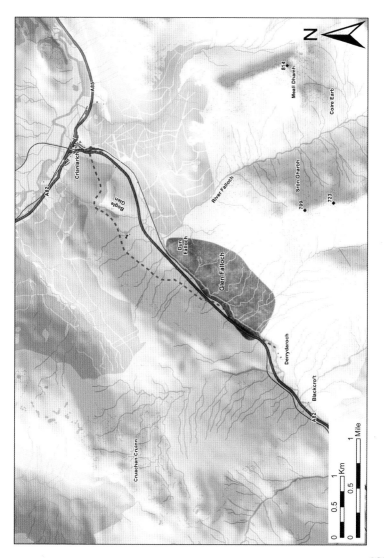

Regenerating pine on Dun Falloch

Travel Notes

The railway station at Crianlarich is only a few kilometres away from the wood. The station tearoom is a well known stop since Victorian times and still has copies of the food and drink orders on the wall that first class customers posted ahead to the old North British Railway in 1901. From the station cross over the A82 to a signposted track that climbs west into a forestry plantation. Join the West Highland Way at Bogle Glen and turn left. The route continues south west along Glen Falloch and under the A82 road, close by the southern limit of the pine trees.

There are guest houses, hotels and a youth hostel in Crianlarich.

Gaelic Place Names

Glen Falloch	hidden glen	Stob Binnein	hill of the anvil
Crianlarich	the low pass		

Glen Falloch along the West Highland Way

BLACK WOOD OF RANNOCH

Map: OS 1:50,000 Sheet No 51
Wood Grid ref: NN 580 560
Access Point: Carie (NN 617 571)

This large pinewood lies on the southern shore of Loch Rannoch deep in the Grampian Mountains and was once the legendary home of thieves and outlaws. The Forestry Commission has had a presence here for over 60 years and since declaring the area a Forest Nature Reserve there has been major restoration of the old pinewood after several decades of adjacent planting with non-native conifers.

This imposing and magnificent wood offers a true wilderness experience as well as catering for a wide range of visitors. The Forestry Commission provides well marked walking and cycling routes and information boards. A challenging but rewarding hill path leads from the Black Wood south through Lairig Ghallabhaich to Glen Lyon.

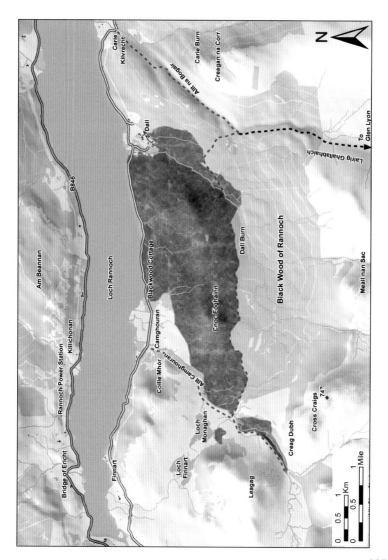

Travel Notes

Rannoch Station is the nearest railway stop. This tiny remote station is no longer staffed and now houses a small tea room and an information display explaining the natural history of the area. From the station the B846 winds its way east across the moors to Loch Rannoch. After about 9 km there is a turning off to the right that continues along the south of the loch. At the Bridge of Gaur the road passes the Rannoch Military Barracks, built in 1746 to house the British troops after the Jacobite uprising of 1745. Another 16 km further on is the Forestry Commission car park at Carie (NN 617 571) where there are information boards and marked paths.

The initial section follows the yellow marked route heading southwest alongside the river, without crossing the bridge then veering right before a shelter to join a forest road. Turn left and continue along this road above the Allt na Bogair for 3 km to a junction. The track leading south is the Old Kirk Road which leads through the Lairig Ghallabhaich to Glen Lyon and the Meggernie pinewood. Turn right here following the yellow marked route past a small lochan which was used in the 19th century as a head of water to float felled trees down to Loch Rannoch. Shortly after, leave the yellow route as it turns off to the right at NN 594 548 and follow the road north west. Ignore the next turning on the right and at the 'T' junction turn left for 300 m and then as the road forks left, go straight on down to the Dall Burn. Stay on the track as it bends left, ignoring the path on the right. At the next junction turn left and follow this track west for 4.5 km to the end of the wood at Allt Camghouran. Cross the footbridge to an estate track and turn left for 1.5 km with the small pinewood remnants below Creag Dubh. Return down the track to the main lochside road at Camghouran and head east back to the carpark at Carie or west towards Rannoch railway station.

The Kilvrecht campsite near Carie is close by the pinewood, providing basic facilities and beside Rannoch station is the Moor of Rannoch Hotel. There are also guest houses in Kinloch Rannoch.

A route from Black Wood of Rannoch to Meggernie

From the Forestry Commission car park at Carie follow the ancient route through the Lairig Ghallabhaich, which is signposted to Glen Lyon. Take the yellow marked route south west along above the Allt na Bogair for 3 km then at the junction turn left to head south. At the southern edge of the forest are the signs of recent massive tree clearing operations where exotic conifers were felled to help regenerate the native pinewood.

There is a fairly rough stony vehicle track steadily climbing uphill across the moorland for 3 km before then making the bumpy descent into Glen Lyon.

At Innerwick the track ends beside a small church and large memorial cairn. From here, take the road west and then turn left to cross the Bridge of Balgie over the River Lyon and the start of Meggernie pinewood.

Gaelic Place Names

Black Wood of Rannoch	black wood of the ferns	Creag Dubh	black crags
Allt na Bogair	burn of the bog	Allt Camghouran	crooked burn of the goat
Kilvrecht	speckled wood	Schiehallion	fairy hill of the Caledonians

MEGGERNIE

Map: OS 1:50,000 Sheet No 51
Wood Grid ref: NN 555 455
Access Point: Bridge of Balgie, Glen Lyon (NN 57 46)

The old wood of Meggernie lies in upper Glen Lyon near Meggernie Castle. Magnificent old trees between 200 and 300 years old cover the slopes on the south shore of the River Lyon and at Croch na Keys. Heavy deer grazing pressure in the last few centuries has limited the growth of young trees but recent conservation efforts are encouraging new growth within protected areas.

Glen Lyon is exceptional for its stunning landscapes and rich historic interest. This is a difficult area to reach by public transport with a long difficult hill trail in from Rannoch Station or Bridge of Orchy. Alternatively there are pleasant walks signed from Bridge of Balgie with panoramic views of the pinewoods. Fortingall Yew, one of the oldest living things in Europe, is 11 kilometres further down the glen beside a famous coaching Inn and there is a cafe at the Bridge of Balgie.

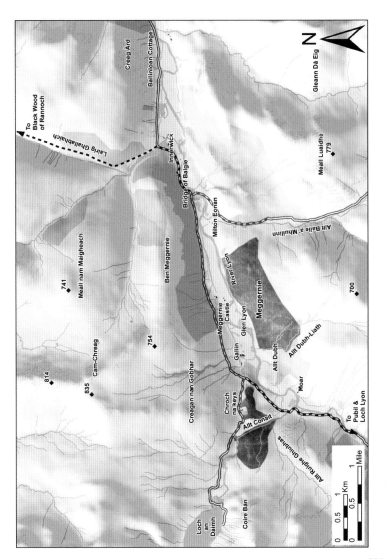

Mature pine above Milton Eonan

Travel Notes

This is a difficult glen to reach by public transport. The nearest train station is at Rannoch to the north and involves travelling through the Black Wood of Rannoch and the Lairig Ghallabhaich to the Bridge of Balgie (see previous chapter for details). Cross the bridge and head south for 1.5 km to view the eastern edge of the wood. Return to the north side of the River Lyon and take the road west past Meggernie Castle. Then after Gallin Farm, take the small road to the right (NN 536 456) that heads west to Loch an Daimh. Around the junction and extending 1 km up the track, is the small pinewood at Croch na Keys.

Recently an alternative route from the Bridge of Orchy train station has been made available by the construction of a new estate track along the south of Loch Lyon. The route is partly a heritage path; an old coffin road used by people in the remote glens to carry their dead to the consecrated ground at Killin. From the station, take the West Highland Way south for 4 km to Auch. Cross the bridge over the Allt Kinglass and turn left to pass under the tall viaduct carrying the West Highland Railway line. Follow the track upstream where there are several deep crossings which have to be waded through. At NN 357 395 the track heads east alongside the Allt a Chuirn and then on the north side of Allt Tarabhan to the west end of Loch Lyon. At the junction turn right to go round the south side of the loch and then join the road past the houses at Pubil and continue on to Meggernie.

Good elevated views of the pinewood can be had by from the viewpoint in the birchwoods on the southern slopes of Ben Meggernie. The start of the trail is at the carpark on the north of the River Lyon, beside the Bridge of Balgie.

There are guest house facilities at Milton Eonan and the Fortingall Hotel further down the glen.

Gaelic Place Names

Meggernie	boggy place	Allt a Chuirn	burn of the cairn
Pubil	the pavilion/tent	Loch an Daimh	loch of the oxen/stag
Allt Tarabhan	burn of the little bulls	Allt Conait	burn of the stream

Safety and access

Scottish Outdoor Access Code

Enjoy Scotland's outdoors responsibly
- take responsibility for your own actions
- respect the interests of other people
- care for the environment.

Particular care should be taken during the deer stalking season, mainly July to October. A pilot scheme 'Heading for the Scottish Hills' provides advice about how to avoid disturbance during deer stalking.

Transport and Accommodation

Suggestions for travel and places to stay or eat are given at the end of each pinewood chapter. These are simply based on the author's own experience and more detailed information can be obtained from:

VisitScotland, Traveline Scotland and Scotrail

Safety

Most of the woods in this guide are remote and located in mountain areas which can have severe weather especially in winter, so expeditions need to be properly planned and participants should be suitably prepared and equipped. Advice on travelling to the Scottish hills is provided by Scottish Natural Heritage.

It is important to have a copy of the relevant OS map or equivalent and ability to use a compass when following any of the routes suggested in this guide to the pinewoods. Nowadays with the advent of mobile phones and global positioning systems it is easy to be lulled into a false sense of security but such equipment is still fallible.

Several of the woods do have good visitor access facilities, including marked trails but even here care is required. Follow any advice and notices provided at these sites.

Dothistroma (Red Band) Needle Blight

A fungal infection (Dothistroma septosporum) harming conifers (principally pine species) has been spreading through parts of Scotland since 2005 and has severely affected Corsican pine and lodgepole pine. Although Scots pine has generally been considered to be of low susceptibility, an increase in the distribution and severity of the disease on this species is now being seen. It is not yet known whether this will lead to widespread mortality or extend significantly into the Caledonian pinewoods.

Rain splash, moist winds and mist spread the disease but infected needles on clothing, footwear, vehicles and machinery can also pose a risk. As this disease is now endemic and is not regulated other than at nurseries, special measures are not required to be taken by visitors – but as a precaution, and in line with good biosecurity principles, it is advisable to remove pine needles from clothing and footwear before entering or leaving woods.

Further information on Dothistroma needle blight can be obtained from the Forestry Commission Scotland. Advice on sensible, routine biosecurity measures is available from the Scottish Government.

Selected Bibliography

Anderson, M.L. 1967. *A History of Scottish Forestry*. Two volumes. Nelson, Edinburgh.

Auldhous, J.R. (Ed.) 1995. *Our Pinewood Heritage*, proceedings of a conference at Culloden Academy, Inverness, Forestry Commission, RSPB, Scottish Natural Heritage, Edinburgh.

Bain, C.G. 1987. *Native Pinewoods in Scotland; A Review 1957-1987*, Royal Society for the Protection of Birds, Sandy.

Bain, C.G. 2013. *The Ancient Pinewoods of Scotland – A Traveller's Guide*, Sandstone Press, Dingwall.

Bunce, R.G.H. and Jeffers, J.N.R. (Eds.) 1975. *Native Pinewoods of Scotland*. Proceedings of Aviemore Symposium, Institute of Terrestrial Ecology, Cambridge,

Jones, A.T. 1999. *The Caledonian Pinewood Inventory of Scotland's Native Scots pine woodlands*. Scottish Forestry 53, pp.237-242.

Laughton Johnston, J. and Balharry, D. 2001. *Beinn Eighe, The Mountain above the Wood*, Birlinn, Edinburgh.

Mason, L.W. Hampson, A. and Edwards, C. (Eds.) 2004. *Managing the Pinewoods of Scotland*. Forestry Commission, Edinburgh.

Peterken, G.F. and Stevenson, A.W. 2004. *A New Dawn for Native Woodland Restoration on the Forestry Commission Estate in Scotland*, Forestry Commission Scotland, Edinburgh.

RSPB, 1993. *Time for pine: a future for Caledonian pinewoods*. Royal Society for the Protection of Birds, Sandy.

Smout, T.C. (Ed.) 1997. *Scottish Woodland History*, Scottish Cultural Press.

Smout, T.C. MacDonald, A.R. and Watson, F. 2005. *A History of the Native Woodlands of Scotland, 1500–1920.* Edinburgh University Press.

Steven, H.M. and Carlisle, A. 1959. *The Native Pinewoods of Scotland*, Oliver and Boyd, Edinburgh. Facsimile edition 1996. Castlepoint Press, Dalbeattie.

Tansley, A.G. 1953. *The British Islands and their vegetation.* Cambridge University Press, Cambridge. pp.253-255 and 444-451.

Wormell, P. 2003. *Pinewoods of the Black Mount*, Countryman, Skipton.

Useful web sites

Cairngorms National Park (www.cairngorms.co.uk)

Forestry Commission Scotland (www.forestry.gov.uk)

Heritage Paths (www.heritagepaths.co.uk)

John Muir Trust (www.jmt.org)

Mountain Bothies Association (www.mountainbothies.org.uk)

Loch Lomond and Trossachs National park (www.lochlomond-trossachs.org)

National Library of Scotland (www.nls.uk)

National Trust for Scotland (www.nts.org.uk)

Native Woodlands Discussion Group (www.nwdg.org.uk)

Scottish Outdoor Access Code (www.outdooraccess-scotland.com)

Ramblers Scotland (www.ramblers.org.uk/scotland)

RSPB (www.rspb.org.uk)

Scottish Natural Heritage (www.snh.gov.uk)

Scottish Wildlife Trust (scottishwildlifetrust.org.uk)

Sustrans Scotland (www.sustrans.org.uk)

Traveline Scotland (www.travelinescotland.com)

Trees for Life (www.treesforlife.org.uk)

Visit Scotland (www.visitscotland.com)

Woodland Trust (www.woodlandtrust.org.uk)

Every effort has been made by the author and publisher to confirm the information in this book is accurate but they accept no responsibility for any loss, injury or inconvenience experienced by any person or persons whilst using this book.

Drawings by Darren Rees

p38	Red Squirrel	p134	Twinflower
p46	Crested Tit	p142	Wood Sandpiper
p62	Pine Marten	p154	Capercaille
p74	Timberman Beetle	p164	Scottish Wildcat
p84	Crossbill	p186	Tree Creeper
p88	Osprey	p190	Golden Eagle
p108	Black-throated Diver